SEMINARS IN
CHIROPRACTIC

A QUARTERLY
SERIES

DIAGNOSTIC DECISION MAKING IN
CLINICAL PRACTICE

VOLUME 2, NUMBER 1, WINTER 1991

SEMINARS IN
CHIROPRACTIC
A QUARTERLY
SERIES

DIAGNOSTIC DECISION MAKING IN
CLINICAL PRACTICE

Jennifer R. Jamison, MBBCH, MSc, MEd
Head, Department of Diagnostic Sciences, Phillip Institute of Technology, Bund(
Victoria, Australia

Series Editors

Dana J. Lawrence, DC
Editor, *Journal of Manipulative and Physiological Therapeutics*
Director, Department of Editorial Review and Publication
Professor, Department of Biomechanics and Chiropractic Technique
The National College of Chiropractic
Lombard, Illinois

Stephen M. Foreman, DC, DABCO
Adjunct Assistant Professor, Department of Diagnosis
Postgraduate Faculty
Los Angeles College of Chiropractic
Whittier, California

WILLIAMS & WILKINS
BALTIMORE · HONG KONG · LONDON · MUNICH
PHILADELPHIA · SAN FRANCISCO · SYDNEY · TOKYO

Editor: Jonathan W. Pine, Jr.
Associate Editor: Marjorie Kidd Keating
Copy Editor: Linda Hansford
Designer: Dan Pfisterer
Illustration Planner: Lorraine Wrzosek
Production Coordinator: Anne G. Seitz
Cover Designer: Dan Pfisterer/Mike Kotarba

Copyright © 1991
Williams & Wilkins
428 East Preston Street
Baltimore, Maryland 21202, USA

ISSN: 1047-1227

All rights reserved. This book is protected by copyright. No part of this book may be reproduced in any form or by any means, including photocopying, or utilized by any information storage and retrieval system without written permission from the copyright owner.

Printed in the United States of America

90 91 92 93 94
1 2 3 4 5 6 7 8 9 10

To

Barbara Cullen

Series Editors' Foreword

SEMINARS IN CHIROPRACTIC has been designed to provide the chiropractic practitioner with clinically and academically relevant information. We will present a mix of material: modern concepts on the topic in question as well as information and procedures that can be directly applied in a busy chiropractic practice. In a sense, this resembles the type of material presented at weekend educational seminars; the best of these programs have always reflected the need to provide practical information.

Each volume (four issues) will cover topics relating to such areas and disciplines as thermography, clinical biomechanics, radiology, research, and clinical diagnosis and management. The most current information needs of the chiropractic profession will be served.

We intend to cover a wide variety of topics; at the same time we hope to develop a reader-driven program that will enable us to be both flexible and responsive. Your comments and suggestions will be gratefully appreciated.

Dana J. Lawrence, DC
Stephen M. Foreman, DC, DABCO

Preface

Diagnostic decision making is the term used to describe problem solving in clinical practice. In primary practice, diagnostic decisions are often requested during the early phases of the health–disease spectrum. Diagnosis at an early stage of disease pathogenesis is particularly challenging as clinical syndromes are often ill-defined and marked overlap occurs between nonspecific signs and symptoms. When the clinician is confronted by such indeterminate clinical presentations, the conscious implementation of recognized diagnostic decision making strategies is helpful in addressing diagnostic dilemmas.

This book considers the processes involved in clinical decision making and discusses how the various principles of diagnostic decision making may be implemented in primary practice. Examples discussed include: the use of probability reasoning in case finding and disease diagnosis, pattern recognition in the differential diagnosis of arthritis, causal thinking as a guide to osteoporosis management, and the minimax strategy as an aid to safe practice in the diagnosis of headaches and chest pain. Diagnosis is the pursuit of an appropriate label for a patient's health status. Selection of a diagnostic label, however, also implies acceptance of an associated management approach; yet confident diagnostic labeling of essential hypertension does not inevitably lead to rational, scientifically validated therapy. Clinical decision making is subject to continuing scrutiny, at both the cognitive and the clinical practice level.

Jennifer R. Jamison, MBBCH, MSc, MEd

Contents

Series Editors' Foreword vii
Preface ix

1 **Clinical Decision Making: Art or Science?** 1

2 **Probability Reasoning: Quantifying Clinical Judgment?** 17

3 **Pattern Recognition: A Signpost to Management** 37

4 **Causal Diagnosis: Unproven Pathogenesis, Uncertain Intervention** 55

5 **Diagnostic Labeling: A Camouflage for Ignorance?** 89

6 **Safe Health Care: The Product of Optimal Clinical Decision Making** 111

Index 129

1

Clinical Decision Making: Art or Science?

Decision making lies at the very heart of the art and science of health care. While the science of clinical practice suggests that decision making should be an objective process in which patients with similar problems are offered equivalent management, the art of health care suggests that individuals respond uniquely. Clinical decision making is based on the accurate collection of patient data, the integration of such data in a pattern congruent with accepted disease pathogenesis, and the comparison of such integrated data with contemporary health management concepts. Clinical decision making is essentially a process whereby the practitioner seeks to compare the clinical data derived from each patient with the gestalt of his or her clinical practice experience and textbook knowledge of health matters. Scientific knowledge is deemed worthy of particular emphasis in primary practice, a notion congruent with the belief that a major constraint of what counts as knowledge in society is the amenability of information to objective assessment (1). "The superiority of scientific knowledge is that it gives advantages of prediction and control" (2). Scientific knowledge works within a framework of specific criteria; it employs operationalized definitions. Adequate definition of criteria is a prerequisite to standardized practice, to comparable clinical decision making, and to evaluating the effectiveness of clinical intervention.

Although scientific knowledge, in view of its objectivity, is considered more likely to predictably ensure clinical success, scientific knowledge is not inevitably or even necessarily value-free. Three distinguishable approaches to the nature of scientific knowledge include (3): the subjective approach, which views scientific knowledge as a special set of beliefs held by individual scientists; the consensus approach, in which beliefs are subservient to their scientific community; and the objectivist view of science, in which scientific theories are deemed to enjoy an autonomous existence independent of their proponents. Both the subjective and the consensus approach are value-laden; the objective approach claims to be value-free.

The subjective approach blatantly permits consideration of beliefs that may be logical but irrational; the consensus approach, which incorporates the commonsense knowledge of everyday health care and is believed to mirror the real conditions of existence, is a more covertly value-laden form of science. In theory, consensus conferences are an attempt to bridge the gap between research and practice (4); in practice, they may constitute well-orchestrated attempts to exert psychological and political pressure by creating a forum for special pleading (5). Issues supported or rejected at consensus conferences are inevitably influenced by the background of participants (6). Nonetheless, expert consensus is regarded as an alternative to scientific certainty and has been used to address a number of disease questions. By 1979, dietary-linked disease issues that had been subject to consensus conference "resolution" included: dietary cholesterol and arteriosclerosis; saturated and unsaturated fats and arteriosclerosis; carbohydrate and sucrose and arteriosclerosis, diabetes, and dental caries; alcohol consumption and arteriosclerosis and liver disease; dietary sodium and hypertension; and excess energy ingestion and obesity, hypertension, diabetes, and arteriosclerosis (7).

It was a consensus conference that concluded that elevation of blood cholesterol is a major cause of coronary artery disease and that lowering raised blood cholesterol will reduce the risk of heart attacks attributable to this cause. This conclusion was expanded to include the concept that appropriate dietary changes would successfully reduce blood cholesterol levels (8). Such a conclusion was made despite the dietary evidence falling short of proof:

> Proof of a causal relation between diet, plasma cholesterol level, and ischaemic heart disease could only be obtained with a randomized and double-blind trial comparing two diets. No completely convincing dietary study for the prevention of heart disease has been published and it is unlikely that the perfect study will ever be conducted: the number of patients needed would be too large; the cost would be too high; dietary studies cannot be double-blind; and it is impossible to prevent a control population from changing its habits (9).

> The panel evidently leaned heavily on epidemiological evidence and public health consideration rather than on clinical viewpoints and data obtained through studies in man (10).

Implementation of such consensus conclusions, based largely on current scientific evidence, does facilitate some standardization of clinical practice in conventional health care, and appropriately structured, the consensus process does provide a means of reviewing research and practice information. Ideally, consensus conference enunciations should be graded from firm recommendations based on rigorous research evidence to tentative guidelines representing inadequately substantiated belief (11). In the absence of rigorous scientific standards, distortions and delusions are inevitable. Like scientifically sound pronouncements, distortions and delusions arising from consensus proceedings ultimately achieve acceptance in clinical practice and incorporation into the nosological (textbook) knowledge of health care.

In contrast to the subjective and the consensus approach to clinical care, the objective perspective suggests that some "science" may be value-free. The implementation in clinical care of interventions assessed by such autonomous scientific pursuits is most likely to ensure predictable results. This autonomous model of science, which most closely resembles the biomedical model, accounts for medical success in disease care. Unfortunately, the results of investigations can be contradictory, given the nuances of research design and analysis; in such circumstances the weight of expert opinion may be called upon to resolve issues better addressed by further investigation. Scientific knowledge, while constituting valid knowledge for clinical information, does not necessarily exclusively suffice as relevant practice knowledge. Clinical practice is a sociocultural pursuit, and different forms of knowledge are pertinent to health care.

Even when employing science, knowledge currently considered valid may not be so regarded at some future date. Therapeutic regimens once considered acceptable can be rejected later; diagnostic criteria acceptable in the 20th century may undergo drastic modification in the 21st. Popper (12) emphasizes that "scientific method uses falsifiability rather than verification as a criterion of demarcation." As such, scientific knowledge is never complete and, as long as research continues, will never be stationary. Science also fails to generate truth; it merely generates more data evaluated by an approved methodology. "Science differs from commonsense knowledge, not in the elimination of preconceptions, but in the precision with which some of these preconceptions are formulated and the detail with which they are used to guide observation" (13). In "normal science" the accepted paradigm influences not only problem selection, but also the methods whereby relevant data are identified, collected, and interpreted. Paradigms guide research by direct modeling and abstracted rules. Kuhn (14,15) defines normal science as research based on past scientific achievements and acknowledged by a particular scientific community as forming the foundations of its future practice. The three classes of problems addressed by normal science are the determination of significant facts, the matching of facts with theory, and the articulation of theory. These problems need to be addressed when seeking to make rational clinical decisions.

In data collection, the patient's signs and symptoms, along with laboratory and radiological information, are used to provide a picture of the patient's clinical presentation. The matching of these facts with currently accepted theories about disease etiology, pathogenesis, and presentation provides a clinical picture in which the significance of various findings in the patient can be assessed relative to that health profession's experience of that condition. Individual case studies, when expanded to clinical trials, can in turn contribute to the further articulation of theory as this pertains to the diagnosis and management of diverse conditions.

As implementation of scientific knowledge has been found the most consistent predictor of outcome, the biomedical model of health care has achieved prominence in disease management. Using knowledge based on the biomedical model, medicine has achieved some reduction in the uncertainty associated with disease pathogenesis, thereby gaining for its adherents a perceptable measure of clinical control. With its deterministic view of health and disease, this model is based on the principles of dualism and reductionism (16). Dualism predicates the separation of mind from body; reductionism assumes that an understanding of a complex entity is best achieved by identifying and analyzing its parts. This scientific framework has facilitated substantial gains in disease care. Disease syndromes are the intellectual constructs that have resulted from clinical observation and theory building. The science of clinical medicine is based on precise observations and theories of causation, decision making, and proof (17). Its methodology allows testing of hypotheses in the real world; while absolute truth can never be proven, falsifying evidence can be disclosed and theory refuted. The basic scientific discipline of the biomedical model is molecular biology; its research model arises from the agricultural-botany paradigm. According to the biomedical model, rational therapy derives from a knowledge of the etiology and pathogenesis of a disease and the clinical responses encountered on attempting to modify the natural history of that disease. Effective clinical management is based on a scientific understanding of the causes and mechanisms underlying the development of a condition, as well as an objective analysis of the clinical response(s) to a hypothesized intervention. The biomedical model is ideal for disease care; it has implications for patient care but has little relevance to "health" care. The clinical expression of the biomedical model is "curing" rather than "caring" medicine (5). The biomedical model largely fails to recognize that medical science and clinical practice are nonidentical—an oversight that becomes particularly apparent at the patient–practitioner interface of primary practice.

The inadequacy of the biomedical model as a framework for primary practice becomes apparent when one considers that primary care must encompass most of the patient's first-contact problems (19). At least one study of general medical practice has found that half of primary medical care encounters are with persons devoid of detectable organic pathology; the pathology present in the remaining persons is self-limiting in seven of every ten cases (20)! According to this study, the biomedical model with its pathological orientation is required for some 15% of primary care clinical consultations. The need for a new model of health care in primary practice is further emphasized by the finding that a specific diagnosis is made for approximately one in five cases of abdominal pain, one in three cases of headache, and one in two cases of chest pains presenting in general practice (21). These findings are not unchallenged; a Canadian study of family practice concluded that, given a classification system that categorized patient problems as physical, psychosocial, lifestyle, or diseases of social impact, some 85% of clinical presentations could be incorporated in the physical grouping (22). Actual family practice time is certainly predominantly dedicated to routine checkups and treatment of respiratory diseases. Although the conflicting findings of these studies may be explained by the different populations studied, they may also be attributable to discrepancies in clinical paradigms and the previous experience of the practitioners surveyed. What these studies do confirm is that effective primary practice requires training in aspects of health care encompassing more than disease diagnosis and treatment. This is acknowledged in family medicine, the medical specialty version of primary practice (23)—a specialty that has been described as having a commitment to people rather than a specialized body of knowledge (24), and which regards the meeting of social need as a higher priority than adherence to an esoteric paradigm of practice.

Health care at the primary level cannot function within the narrow confines of health as the absence of disease; it rests upon the comprehensive definition of health as physical, psychological, and social well-being as proposed by the World Health Organization. Such a perspective is expressed in Engel's (16,25) call for a biopsychosocial model. This model, while adherent to a cause–effect philosophy, does allow for consideration of behavioral variables. Such a model could permit the recognition of lifestyle both as a cause of disease and as a consequence of influences such as social stress, personality, and social support systems (26). Only so comprehensive a model can hope to simultaneously accommodate the notion of disease, an organic state; illness, an individual's psychophysiological or personal perception of not feeling well; and sickness, a social phenomenon (27). This health model would permit consideration of not only the sign and symptom manifestations of disease, but also the patient's feelings/illness manifestations and behavior/sickness manifestations (28). The desirability, indeed necessity, of an expanded model of health care has long been recognized by the chiropractic profession. Chiropractors have achieved clinical success by implementing in their practice—along with such concepts as the conservative ethic (do no harm), self-healing (innate intelligence), the strategic role of the nervous system (subluxation/adjustments), and prevention—a holistic approach to patient care (29).

It is well recognized that clinical practice is distinct from medical science (30). "Just as science is not practice, practice is not merely applied science. Different purposes lead to different activities, associations, information and codes of behavior" (4). Freidson (31) maintains that "the practitioner is quite different from the scientist" and vindicates his stance by evoking clinicians' commitment to action (rather than knowledge), their faith in their work, their pragmatism, and their emphasis on clinical indeterminacy. While clinical uncertainty may appear incongruous in juxtaposition with scientific predictability, confidence in personal experience rather than accumulated scientific "book" knowledge may be even more anomalous. Depending on the patient's presenting complaint, clinical success may be no less attainable using the art than the science of medicine. Chiropractic's noted clinical achievements and acknowledged paucity of scientific evaluation provide adequate testimony that clinical success can, on occasion, be achieved using therapeutic interventions that lack scientific validation. Furthermore, if science is not absolute truth but only an approximation of truth, and some forms of scientific investigation are acknowledged to be value-laden, then primary practice cannot on this basis justify outright rejection of patient belief systems. On the contrary, a cognizance of patient beliefs needs to be acknowledged as an integral part of clinical decision making. In the final analysis, it is clinical effectiveness, rather than truth or even science, that is the predominant pursuit of clinical practice.

Regardless of whether a biomedical or a holistic model of health care is implemented in the pursuit of clinical effectiveness, clinical data have three separate roles in clinical reasoning (32): to suggest a group of causative factors or associated diseases in diagnosis; to indicate comparable patient groups in prognosis and therapeutic strategy; and to target for treatment those groups of clinical manifestations considered distressing by the patient. For each of these clinical activities, the intellectual constructions and systems of classification differ. In each of these cases, the precision and accuracy with which clinical data are collected and recorded form the foundation of competent clinical decision making. The precision of clinical findings may be ascertained by monitoring and limiting or excluding intra- and interobserver variation. The accuracy of observations and recorded data may be determined by checking findings using different investigations. A clear definition of terminology and symptomatology further enhances reliability. Data relevant to clinical decision making according to the biomedical model include demographic information, the age, sex, and occupation of the patient; a detailed description of the subjective symptoms, the presenting complaint; the recording and confirmation of the patient's objective symptoms; the detection of other physical signs; and the ordering or performance of relevant laboratory or radiological investigations. Socioeconomic, psychoemotional, and cultural data should also be collected prior to decision making, using the more comprehensive holistic or biopsychosocial model of health care. Qualitative data may be dichotomously recorded as present or absent, semiquantitative data may be described on a somewhat arbitrary scale, and quantitative data may be precisely assessed and recorded on an interval scale. Decisions made using an incomplete or incorrect data base may well be erroneous, the seriousness of omissions and/or inaccuracies only becoming apparent retrospectively. Decisions should be made on the basis of data collected; data rather than the clinician's interpretation of findings should consequently form the essence of the clinical record.

The collected data are then integrated in the light of the clinician's previous experience and nosological or textbook knowledge of disease. Diseases may be classified according to symptoms, anatomically defined diseases, syndromes, and causally defined diseases (33).

1. Symptoms. Diagnoses target a predominant symptom and emphasize the patient's complaint, e.g., fatigue. A sign may also be used as the basis of diagnosis, as in the case of essential hypertension. This approach to diagnosis may be expanded to provide a problem-oriented record (34). Problem-oriented decision making is aimed at meeting the patient's perceived and assessed needs. This is a legitimate approach in holistic primary care circles; primary practice need not be confined within a specific paradigm as it is a specialty meeting a functionally defined social need (35). Certainly, reasoning that results in a listing of patient problems rather than a tentative diagnosis does serve to ensure that patients will receive interventions that they regard as relevant.

2. Anatomically defined diseases. Headache or chest pain may be preliminary resting points in diagnostic decision making. Anatomically defined disease may be a valuable means for patients to summarize their presenting complaint, but for clinicians such nonspecific statements suffice only as a preliminary diagnostic comment in the interim between patient consultation and the receipt of laboratory and radiological results. In contemporary health

care, a "diagnosis" presented as an unqualified anatomically defined disease statement constitutes little more than an admission of clinical uncertainty. In primary practice, clinical uncertainty may constitute a particular problem. Patients may present to their primary practitioners at a particularly early stage of disease pathogenesis. When required to make clinical decisions in the face of ill-defined clinical manifestations, the practitioner may opt to wait until the disease pattern becomes more clearly defined or to intervene, making a tentative diagnosis based on probability reasoning. Probabilistic diagnostic reasoning can be implemented at all stages in the natural history of disease; it is a particularly useful strategy in case finding and population screening.

3. Syndromes. Frequently observed patterns of signs and symptoms are correlated and labeled as particular disease entities. Rheumatic fever is diagnosed when two major or one major and two minor manifestations are present. Major manifestations include carditis, polyarthritis (usually migratory), erythema marginatum, subcutaneous nodules, and Sydenham's chorea; minor manifestations are fever, arthralgia, ECG changes with a prolonged P–R interval, a history of previous rheumatic fever, the presence of a raised antistreptolysin-O titer and/or evidence of an inflammatory process: raised erythrocyte sedimentation rate, leukocytosis, or elevated C-reactive protein. Rheumatoid arthritis is another condition in which the diagnosis depends on the quantity and quality of clinical signs and/or symptoms present. Pattern recognition is an accepted diagnostic strategy.

4. Causally defined diseases. Deductive or causal diagnosis emphasizes the importance of disease pathogenesis. It exemplifies implementation of the scientific hypothetico-deductive approach in clinical decision making. Hypothetico-deductive reasoning employs the logic underlying both biological and agricultural science traditions. The categorization of conditions along etiopathological lines has profound implications for intervention. The one cause–one disease perspective has largely been discredited. Instead, a more holistic approach has emerged that recognizes the multifactorial nature of disease etiology and attempts to address each of the contributing factors. Disease determinants have been classified as: necessary, the disease does not occur in the absence of this variable; sufficient, this variable is alone capable of initiating the disorder; and contributory. The presence of diverse variables may increase the likelihood or probability of manifest disease.

Distinct trends in diagnosis may be identified, but in practice an integrated approach using elements of probability reasoning, pattern recognition, and causal thinking frequently emerges. In all cases the data collected are the same. Probabilistic reasoning, however, by emphasizing the sensitivity and specificity of data collection techniques, increases the likelihood that syndrome patterns are accurately described. Causal thinking is a prerequisite to specific rather than palliative intervention. Regardless of the diagnostic strategy emphasized, the ultimate goal of diagnostic reasoning is safe, effective intervention at minimal patient cost. Prioritization of diagnostic hypotheses according to the principles of probability, seriousness, treatability, and novelty attempts to address this goal (36). More precisely, this goal incorporates several, sometimes conflicting, objectives.

1. Investigating and treating the most probable diagnosis/disease.
2. Selecting a strategy that will ensure the early exclusion of serious disease. The "minimax" strategy is to minimize the maximum loss. It seeks to exclude serious, though improbable diagnoses, so that any condition with a poor prognosis may be treated early. The minimax strategist seeks to avoid any false-negative results. This may be viewed as safe health care; when taken to the extreme it constitutes defensive medicine.
3. Providing a service with a favorable cost–benefit ratio. Maximization of expected utility involves weighing the probability of the disease being missed against the benefits and risks accruing from exposing the patient to various diagnostic tests and/or interventions. In management, it confronts the situation in which an intervention with possible but unproven benefit and recognized side effect carries a cost-benefit ratio that potentially approximates infinity.
4. Providing effective management of the patient's condition. The effectiveness of therapeutic intervention may be presumed on the basis of personal clinical experience, biological plausibility, epidemiological evidence, and controlled clinical trials.

Despite traditional chiropractic practice having emphasized a pragmatic approach to clinical practice, the need for controlled clinical research in chiropractic practice is now readily acknowledged. Anecdotes, while providing ideas for further investigation, cannot be deemed a suitable basis for professionally justifiable clinical intervention. Biological plausibility, although congruent with the logic and reality of disease mechanisms, cannot offer validation for the clinical reality of the patient–practitioner encounter. Epidemiological surveys, while contributing to nosological knowledge by suggesting associations, cannot confirm cause–effect relationships. Key techniques in epidemiology are comparison and contrast; interventions based on epidemiological evidence rely largely on circumstantial evidence of a cause–effect relationship. Epidemiology, the study in human populations of the frequency and distribution of diseases in terms of time, place, and personal characteristics, is nonetheless an important source for clarifying the early phases of disease pathogenesis. Extrapolation of conclusions derived from epidemiological studies of dissimilar population groups to individual patients, however, can be just as erroneous as direct extrapolation of the results of animal experiments to human health care.

The types of clinical traits, in order of increasing validity, clinical practice relevance, and logistic and ethical barriers, are retrospective case control or cross-sectional surveys, observational cohort or prospective longitudinal studies, and experimental randomized clinical trials. Only controlled clinical trials can claim scientific justification for the effectiveness of therapeutic programs. The controlled clinical trial may constitute the gold standard for clinical decision making, but anecdote, biological plausibility, and epidemiological findings are important elements of the research process. The idea that evolves into scientific knowledge in clinical practice may start with anecdote and, once adequate criteria have been defined, emerge as a case study, later to be tested in one or more clinical trials. Despite the acknowledged difficulties in implementing controlled clinical trials, the Lugano conference concluded that progress in medicine depends on controlled clinical trials and that the alternative of practicing in the face of constant uncertainty outweighs any possible disadvantages associated with the conduct of such trials (37). "Clinical practice must be based on the results of clinical trials, not on theories derived from epidemiological observations" (38).

The randomized clinical trial is accepted as the most reliable method for determining the relative merits of different therapies (39,40); unless the main sources of bias can be avoided and recognized, there is no satisfactory alternative to the randomized clinical trial (40). Random allocation of patients to treatment and control groups is the accepted technique for avoiding bias in the allocation of individuals in clinical trials. Persons in the treatment and control groups should be selected according to the same entry and rejection criteria and subject to the same clinical procedures, except that the control subjects do not actually receive the therapy under investigation. Control subjects receive instead a superficially similar but sham intervention. Control groups are essential prerequisites to therapeutic trials because treatment "responses" may result from therapy, a placebo effect, observer bias, and spontaneous remission. Using methods reminiscent of the agricultural-botany paradigm, clinical data are randomized and results analyzed using statistical measures (41,42).

An inherent difficulty in this approach is the potentially fallacious assumption that measures on members of the research and/or therapeutic population form a gaussian distribution, as would repeated measures on the same sample. Application of the theoretical gaussian or "normal" distribution to clinical practice implies that the clinical indices of "healthy" people would fall within a range of the mean plus or minus two standard deviations of the distribution of a carefully defined reference population. As the gaussian curve extends to infinity on both tails, this suggests that some clinical findings are on the minus side of zero.

A second clinical nonsense inherent in the application of gaussian science to patient care is the logical extension of the statistical designation that 2.5% of diagnostic test results are abnormal to imply that all diseases have the same frequency. Bayesian strategy, in providing a computational technique for diagnostic probability, uses the grammar of mathematics and supplies a name for, but not an explanation of, clinical evidence. Clinical decision making, while seeking to establish itself as an objective process, should remain "a rational activity of clinical medicine rather than a computational exercise in numerical statistics" (43). Statistical reasoning inappropriately applied may result in clinical nonsense (44). A statistically significant result implies only that the result is unlikely to be attributable to chance; it does not provide information about the magnitude of the biological effect of the intervention. A statistically significant result does not provide information about the therapeutic importance or gain to be derived from a therapy; it does provide information about the likelihood of an intervention having a clinical effect beyond that of chance. Even using randomized controlled clinical trials it is possible to arrive at incorrect conclusions. Statistical analyses of clinical trials, while reducing the likelihood that observed responses are inappropriately deemed cause–effect responses, do recognize type I and type II errors. A type I error occurs when the clinical researcher incorrectly concludes that a therapy achieves a benefit greater than that which can be attributed to chance; type II errors exemplify the erroneous rejection of an efficacious therapy. Type I errors may be manifestations of the placebo effect; type II errors may exemplify the "tomato effect" (45,46).

Despite extensive research funding, nosological knowledge remains incomplete. The scientific strategy used to address unanswered clinical dilemmas is more science, i.e., further investigation. In the absence of a definitive clinical trial, diverse epidemiological methods, although individually inconclusive, may collectively represent a hierarchy of truth (40). Individual practitioners may attempt to address this dilemma by categorizing information according to the Canadian Task Force's (47) ranking of interventions. The Task Force developed a grading system based on the quality of the evidence obtained:

I. Evidence obtained on at least one properly randomized trial.
II-1. Evidence obtained from well-designed cohort or case-control analytical studies, preferably from more than one research center.
II-2. Evidence obtained from a comparison between times or places, with or without intervention; dramatic results on uncontrolled experiments are included in this category.
III. Opinions of respected authorities based on clinical experience, descriptive studies, or reports of expert committees.

By selecting data concerning nutrition from refereed journals, checking the authors' qualifications and research records, analyzing the research design, and confirming that the data warrant the conclusions drawn, practitioners can screen clinical information and determine whether this meets the prerequisites for use in clinical practice.

Experts have individually and collectively also attempted to address the dilemma of inadequate research and contradictory research findings. Authorities have sought resolution by means of innovations in data analysis (48,49) and consensus amongst experts. In both cases, data analysis and/or interpretation may be flawed because of sociopolitical factors. Metaanalyses of experimental epidemiological data have been shown to approximate the results of large well-conducted trials (50). But while results obtained by metaanalyses are reasonably reproducible (51), they may be invalid because of publication bias. When contradictory results are challenged by more research, it may become possible to identify and resolve methodological defects; consensus conferences may impede such scientific progress (52). If disputes arising from conflicting evidence are resolved by a political rather than a scientific process, dogma is encouraged and scientific flaws are not addressed. Public policy can be decided by consensus decisions; consensus decisions cannot resolve scientific controversies (53).

In addition to diagnosing overt disease, primary practitioners are called upon to prevent disease and promote health. Once again, primary practitioners are challenged by the management of patients whose health status approximates the health rather than the disease pole of the health–disease spectrum. Disease prevention depends on a clear understanding of the natural history of disease. Nosological knowledge about the early natural history of disease is derived from epidemiological studies. An advance in the comprehension of variables influencing the early evolution of disease requires a naturalistic field study orientation rather than a laboratory or bedside experimental approach. It requires naturalistic observations and a more inductive logic than the hypothetico-deductive reasoning of clinical inquiry.

Epidemiology, the research tradition that studies the distribution and determinants of disease in human populations, is adept at demonstrating the multifactorial nature of disease. Logistic difficulties, economic costs, and an increased emphasis on prevention have encouraged a contemporary trend in clinical research toward increased reliance on and acceptance of quantitative epidemiological methods (54). Despite their increasing popularity, epidemiological studies may be fraught with methodological problems (55). Feinstein (56), a clinical epidemiologist, illustrating his argument with cogent examples, cautions against the uncritical reading of epidemiological research reports owing to problems often overlooked at such fundamental levels of research activity as data collection and analysis. He explicitly recommends that "Before we reach editorial decisions about massive changes in national diets, lifestyles, and rigorous regulation of various alleged industrial hazards, we might first establish better scientific standards for getting credible news, and for deciding when the available news is credible enough to warrant the editorials." Epidemiological surveys or ecological association studies identify statistically significant correlations in population groups; but population correlations may have no biological significance for particular individuals. Conclusions are based on probabilities and on traditional, but arbitrary, definitions of the power of chance; on time trends in aggregates of data, rather than on trends in the individual. Application of such findings to individuals requires further investigation; even when epidemiological association leads to scientific consensus, the lack of a causal relationship for a given condition results in different groups proposing different intervention strategies (57). Epidemiological statistics may show an indisputable correlation, but given the nature of the research program, a cause–effect relationship is difficult to establish. Preventive measures for many disorders of multifactorial etiology, including coronary heart disease, pulmonary emphysema, various cancers, hypertension, and obesity, are seldom supported by direct evidence that can withstand scientific scrutiny (58). Chiropractors need to be aware that the scientific basis of generally accepted preventive interventions in apparently healthy people is not beyond reproach.

When seeking to prevent disease in individual patients by applying epidemiological information in the clinical situation, one must ensure that apparently healthy people are neither incorrectly labeled nor inappropriately managed. Case finding or population screening relies on the collection of data using sensitive tests. A sensitive test is one that is unlikely to give a false-negative result. Its usefulness as a screening tool is that it recognizes when a disease variable is truly absent. Persons who have rendered positive results on screening tests are not usually diagnosed on the basis of a sensitive screening test; they are subjected to diagnostic tests with a high specificity. Diagnostic labeling, when used as a therapeutic index, should be based on tests having a high degree of specificity, i.e., tests that are only positive when disease is truly present. In contrast to sensitive tests, high-specificity tests are selected for their low incidence of false-positive results. The predictive value of a test, in addition to its sensitivity and specificity, is influenced by the prevalence of the disorder in the population under investigation.

Clinical decision making in primary practice is further complicated by its health promotion goals. The health promotion approach focuses on augmenting the wellness of the client; it can be clearly differentiated from the avoidance or protective behaviors characteristic of preventive interventions (59). Attempts to implement an inclusive and value-laden conceptualization of health are hindered by the elusiveness of precise health-promoting variables. Evans (60) specifically cautions that establishing proof of the causation of health will be more difficult and challenging than determining the causation of disease. Various health status instruments are available. They vary according to the manner in which health is defined and the function (public information, administration, or medical science) that the indicator is intended to serve. While benefits are postulated to arise from implementing functional health status assessments, any objective contribution in primary practice remains to be demonstrated (61).

Clinical decision making is essentially a problem solving process (62). Inherent in any patient–practitioner encounter are elements of problem definition, solution postulation, prioritization of possible solutions, appropriate intervention selection, implementation of a selected intervention strategy, and evaluation of the outcome. In the clinical encounter, data collection and integration and hypothesis generation aim at initially determining:

- Why the patient has presented at this time.
- The nature of the patient's overt complaints(s).
- The presence of any covert complaints.

This information is then further analyzed specifically:

- To exclude the possibility of serious organic disease.
- To identify variables amenable to intervention, whether at the level of therapy, disease prevention, or health promotion.

Based on a sound diagnosis, therapeutic decisions are then made with regard to:

- Urgency. Immediate intervention may be required or therapy may be delayed.
- Patient care. Management of the overt condition may best be accomplished by treatment, patient referral, or interprofessional consultation.
- Management of the presenting complaint so as to alleviate symptoms, prevent complications, and avoid recurrences.
- Long-term patient health promotion and disease prevention.
- Optimal utilization of the patient's health resources, taking into account the patient's cultural and religious beliefs, health values, economic circumstances, psychological coping mechanisms, and social and family support systems.

The chosen intervention strategies are then monitored for:

- The patient's objective physical response to therapy.
- Professional criteria of what constitutes adequate health care.
- The patient's and clinician's perception of what constitutes a successful outcome.

Traditionally, clinical teaching has emphasized the transmission of recommendations and algorithms based on expert or consensus opinion and clinical trials and/or case studies. Such an approach has served to encourage intuitive judgments (63). Heuristics or intuitive procedures can, however, lead to cognitive biases or systematic errors in judgment. Practitioners may intuitively and erroneously equate the ease of remembering specific instances with the probability of their occurrence; human information processing is relatively inept at using "negative evidence" (normal findings), and clinicians, like nonclinicians, tend to seek only evidence that can be used to confirm hypotheses (64). Decision analysis techniques, with their sophisticated computer program backing, can overcome a number of the limitations associated with clinical decision making (65), but the art of patient care should not be neglected. Feinstein (68) cautions that patient care is not necessarily well served by the belief that clinical thinking is merely the application of fundamental concepts derived from other disciplines.

In theory, the practitioner brings the total knowledge and expertise of his or her profession to bear on the patient's health problem; in practice, the clinician offers assistance within the confines of his or her own clinical experience and somewhat limited nosological knowledge. In clinical decision making, principles derived from the biomedical sciences are helpful in explaining disease pathogenesis. Information derived from the clinical use of randomized trials, bayesian decisions, multivariant statistics, and cost–benefit analyses contributes to rational managerial decision making. These science-based tools provide useful aids for refining clinical decisions; they complement but do not displace patient–practitioner interaction. Ultimately, clinical decisions are made not only in the light of the natural history of disease, but also in the dynamic context of patient values, beliefs, and socioeconomic circumstances. An awareness of the discrete elements involved in clinical decision making helps individual practitioners to maximize their ability to offer a relevant, cost-effective, and safe form of health care.

References

1. Young MFD: An approach to the study of curricula as socially organized knowledge. In Young MFD (ed): *Knowledge and Control*. London, Collier Macmillan, 1971, pp 19–46.
2. Young A: The discourse on stress and the reproduction of conventional knowledge. *Soc Sci Med* 14B:133–146, 1980.
3. Chalmers AF: *What is This Thing Called Science?* St. Lucia, University of Queensland Press, 1978, pp 93–112, 123.
4. Greer AL: The two cultures of biomedicine: can there be consensus? *JAMA* 258:2739–2740, 1987.
5. Oliver MF: Consensus or nonconsensus conferences on coronary heart disease. *Lancet* 1:1087–1089, 1985.
6. Blum AL, Chalmers TC, Deutsch E, Koch-Weser J, Langman M, Rosen A, Tygstrup N,

Zentgraf R: Differing attitudes of industry and academia towards controlled clinical trials. *Eur J Clin Invest* 16:455–460, 1986.
7. Ahrens EH: Introduction to the report of the task force on the evidence of diet related diseases. *Am J Clin Nutr* 32:2627–2631, 1979.
8. Consensus Conference. Lowering blood cholesterol to prevent heart disease. *JAMA* 253:2080–2086, 1985.
9. Editorial. Primary prevention of ischaemic heart disease with lipid-lowering drugs. *Lancet* 1:333–334, 1988.
10. Ahrens EH: The diet-heart question in 1985: has it really been settled? *Lancet* 1:1085–1087, 1985.
11. Lomas J, Anderson G, Murray E, Vayda E, Roberts R, MacKinnon B: The role of evidence in the consensus process. *JAMA* 259:3001–3005, 1988.
12. Popper KR. *The Logic of Scientific Discovery.* London, Hutchinson, 1975, p 41.
13. Mulkay M: *Science and the Sociology of Knowledge.* London, George Allen & Unwin, 1979, p 47.
14. Kuhn TB: *The Essential Tension.* Chicago, University of Chicago Press, 1977, pp 293–319.
15. Kuhn TB: The structure of scientific revolutions. In Neurath O, Carnap R, Morris C (eds): *International Encyclopedia of Unified Science.* Chicago, University of Chicago Press, 1970, vol 2(2).
16. Engel GL: The biomedical model: a procrustean bed? *Man Med* 4:257–275, 1979.
17. McWhinney IR: Family medicine as a science. *J Fam Prac* 7:53–58, 1978.
18. Gordon D: Curing medicine and caring medicine. *Aust Fam Physician* 7:779–784, 1978.
19. Colwill JM: The shifting emphasis in the delivery of health care—viewpoint of a physician. In *Primary Health Care—Everybody's Business.* Council of Hospital and Related Institutional Nursing Services Publications (No. 20-1482). New York, National League for Nursing, 1973, pp 9–19.
20. Carmichael LP: The relational model: a paradigm for family practice. *J Fla Med Assoc* 67:860–862, 1980.
21. Graham N: Our patients are ready for a new model. *Aust Fam Physician* 13:796, 1984.
22. Warrington AM, Ponesse DJ, Hunter ME, Grant DA, Grasset AV, Gray DW, Hayward CD, Long BF, Morrison GEC, Sutherland D: What do family physicians see in practice? *Can Med Assoc J* 117:354–356, 1977.
23. Shapiro J: The maturing of family medicine; challenges to behavioural science. *J Fam Pract* 26:583–584, 1988.
24. McWhinney IR: Family medicine in perspective. *N Engl J Med* 293:176–181, 1975.
25. Engel GL: The need for a new medical model: a challenge for biomedicine. *Science* 196:129–135, 1977.
26. Badura B: Life-style and health: some remarks on different viewpoints. *Soc Sci Med* 19:341–347, 1984.
27. Erde EL: Philosophical considerations regarding defining "health," "disease" etc. and their bearing on medical practice. *Ethics Sci Med* 6:31–48, 1979.
28. Barrand J: A model of health. *Aust Fam Physician* 14:1302–1307, 1985.
29. Keating JC: Five contributions to a philosophy of the science of chiropractic. *J Manipulative Physiol Ther* 10:25–29, 1987.
30. Ruane TJ: Paradigms lost: a central dilemma for academic family practice. *J Fam Pract* 27:133–135, 1988.
31. Freidson E: *Profession of Medicine.* New York, Harper & Row, 1970, p 169.
32. Feinstein AR: Scientific methodology in clinical medicine. II. Classification of human disease by clinical behavior. *Ann Intern Med* 61:757–781, 1964.
33. Wulff HR: *Rational Diagnosis and Treatment.* Oxford, Blackwell Scientific Publications, 1976.
34. Weed LL: *Medical Records, Medical Education and Patient Care. The Problem-Oriented Record as a Basic Tool.* Cleveland, Western Reserve University, 1969.
35. Phillips TJ: Disciplines, specialties and paradigms. *J Fam Pract* 27:139–141, 1988.
36. Elstein AS, Kagan N, Shulman LS, Jason H, Loupe MJ: Methods and theory in the study of medical inquiry. *J Med Educ* 47:85–92, 1972.
37. Blum AL, Chalmers TC, Deutsch E, Koch-Weser J, Rosen A, Tygstrup N, Zentgraf R: The Lugano statements on controlled clinical trials. *J Int Med Res* 15:2–22, 1987.

38. Hampton JR: Mild hypertension: to treat or not to treat? *Nephron* 47(suppl 1):57–61, 1987.
39. Forbes JF: Highlights in the development of randomized clinical trials. *Med J Aust* 1:159–160, 1981.
40. Editorial. Randomized clinical trials. *Br Med J* 274:1238–1239, 1977.
41. Peto R: Clinical trial methodology. *Biomedicine* (Special Issue) 28:24–36, 1978.
42. Zelen M: Statistical options in clinical trials. *Semin Oncol* 4:441–446, 1977.
43. Feinstein AR: An analysis of diagnostic reasoning. I. The domains and disorders of clinical macrobiology. *Yale Biol Med* 46:212–232, 1973.
44. Jamison JR: The use of inferential statistics in health and disease: a warning. *S Afr Med J* 57:783–785, 1980.
45. Podell RN: The 'tomato effect' in clinical nutrition. *Postgrad Med* 76:49–65, 1984.
46. Goodwin JS, Goodwin JM: The tomato effect. *JAMA* 251:2387–2390, 1984.
47. Canadian Task Force: The periodic health examination. *Can Med Assoc J* 121:1193–1254, 1979.
48. Dyer AR: A method for combining results from several prospective epidemiological studies. *Stat Med* 5:303–317, 1986.
49. Anderson JP, Bush JW, Berry CC: Internal consistency analysis: a method for studying the accuracy of function assessment for health outcome and quality of life evaluation. *J Clin Epidemiol* 41:127–137, 1988.
50. Chalmers TC, Levin H, Sacks HS, Reitman D, Berrier J, Nagalingam R: Meta-analysis of clinical trials as a scientific discipline: I. Control of bias and comparison of large with large cooperative trials. *Stat Med* 6:315–325, 1987.
51. Chalmers TC, Berrier J, Sacks HS, Levin H, Reitman D, Nagalingam R: Meta-analysis of clinical trials as a scientific discipline: replicate variability and comparison of studies that agree and disagree. *Stat Med* 6;733–744, 1987.
52. Feinstein AR: Fraud, distortion, delusion, and consensus: the problems of human and natural deception in epidemiological science. *Am J Med* 84:475–478, 1988.
53. O'Hara-May J: Nutritionists' view on nutrition. *Hum Nutr Appl Nutr* 41A:265–284, 1987.
54. Goldberg RJ: Clinical uses of the epidemiological approach. *Hosp Pract* 18:177–189, 1983.
55. Levitan A, Baglivo J, Kuban KCK, Brown ER, Krishnamoorthy KC: The prophylactic clinical trial as an epidemiological resource. *Controlled Clin Trials* 8:243–254, 1987.
56. Feinstein AR: Scientific standards and epidemiologic methods. *Am J Clin Nutr* 45:1080–1088, 1987.
57. Hitt C: Nutrition and risk reduction. In Faber MM, Reinhardt AM (eds): *Promoting Health through Risk Reduction*. New York, Macmillan, 1982, pp 176–182.
58. Feinstein AR: Scientific methodology in clinical medicine. III. The evaluation of therapeutic response. *Ann Intern Med* 61:944–967, 1964.
59. Chalmers K, Farrell P: Nursing interventions for health promotion. *Nurse Pract* Nov/Dec:63–64, 1983.
60. Evans AS: Causation and disease: a chronological journey. *Am J Epidemiol* 108:249–258, 1978.
61. Bergner M: Functional health assessments: are they ready for use in clinical practice? *J Fam Pract* 23:423–424, 1986.
62. Jamison JR: *Holistic Health Care*. Bundoora, PIT Press, 1984, pp 1–19.
63. Detsky AS: Decision analysis: what's the prognosis? *Ann Intern Med* 106:321–322, 1987.
64. Dawson NV, Arkes HR: Systematic errors in medical decision making: judgment limitations, *J Gen Intern Med* 2:183–187, 1987.
65. Kassirer JP, Moskowitz AJ, Lau J, Pauker SG: Decision analysis: a progress report. *Ann Intern Med* 106:275–291, 1987.
66. Feinstein AR: The intellectual crisis in clinical science: medaled models and muddled mettle. *Perspect Biol Med* 30:215–230, 1987.

2

Probability Reasoning: Quantifying Clinical Judgment?

Probability reasoning underpins clinical decision making at all stages of the health–disease spectrum. Toward the health pole of the spectrum, probability reasoning underlies the periodic health examination; at the disease pole, probability reasoning is used to convert an array of possibilities into a single strong probability. At the health pole, the likely risk that a nondiseased person will develop a particular disease is predicted; at the disease pole, the patient's clinical presentation provides the basis of pattern recognition and the foundation upon which hypothetico-deductive reasoning can select a likely definitive diagnosis from an array of possible diagnoses. Hypothetico-deductive reasoning leading to probability diagnosis is not the only strategy used to reach a definitive diagnosis, but it is the approach most frequently employed by experienced clinicians after initial pattern recognition. Alternative clinical decision making strategies include the stepwise implementation of disease algorithms and screening flow charts, or the exhaustive and exhausting complete physical examination and history taking in diseased patients and multiphasic screening in healthy clients. Regardless of the diagnostic strategy used, the objective is to convert a hypothesized possibility into a highly probable diagnosis.

Clinical data are gathered in order to make a diagnosis, judge the severity of an illness, predict the subsequent clinical course of the condition, estimate the likely response to therapy, and determine the actual response to intervention. While particular reasons for collecting clinical data vary, the overriding criterion that determines whether a piece of information should be sought is its usefulness. The usefulness of a piece of clinical information may be objectively ascertained by the extent to which it contributes to the conversion of a possible working hypothesis into a probable diagnosis. Sackett and coauthors (1) have addressed this issue particularly as it relates to disease diagnosis; Frame and Carlson (2) have emphasized the use of probability and outcome in developing criteria for case finding and health maintenance; and the Canadian Task Force (3) has evolved protocols for periodic health examinations according to the likelihood of covert disease. The whole concept of health promotion and disease prevention rests upon the likelihood that disease will develop given certain circumstances, and clinical intervention can only be justified by the probability that such management improves the patient's health status. Probability reasoning can be construed as an essential strategy in the delivery of efficient health care. It is applicable at all points of the health–disease spectrum.

In any client/patient–practitioner encounter, a number of diverse probabilities interact. The possibility of disease in an individual belonging to a particular population group may be roughly ascertained from mortality and morbidity statistics. In 1987, the ten leading causes of death in the United States were heart diseases (accounting for 35.7% of the total deaths, some 24.1% being attributable to coronary heart disease); cancers (22.4%); strokes (7%); accidental injuries, including motor vehicle accidents (4.4%); chronic obstructive lung disease (3.7%); pneumonia and influenza (3.2%); diabetes mellitus (1.8%); suicide (1.4%); chronic liver disease and cirrhosis (1.2%); and atherosclerosis (1.1%) (4). The crude possibility that the presenting individual may suffer from any of the listed conditions can be refined by categorizing mortality and morbidity statistics in terms of age and sex. Australian mortality statistics show that 26% of deaths in men between the ages of 20 and 59 years are attributable to cancer, 25% to ischemic heart disease, 24% to accidents, and 4% to stroke; in men over the age of 60 years, ischemic heart disease and stroke death rates increase to 33% and 10%, respectively (5). Similar statistical analyses of death rates in 20- to 50-year-old women attribute 44% of deaths to cancer, 14.5% to accidents, 11% to ischemic heart disease, and 6.5% to strokes; after the age of 60 years, ischemic heart disease and stroke death rates rise to 29.5% and 17%, respectively, while overall cancer mortality drops to 20%.

Further personalization of probability data based on nosological knowledge about the natural history of various diseases can be used to focus more clearly on the disease risks of individual patients and clients. The likelihood of covert disease being present in any particular individual can be assessed by an active search for one or more risk factors that have been found to be associated with particular conditions. Males and postmenopausal females are at greatest risk for a heart attack when they smoke, are hyperlipidemic, are hypertensive, have a family history of ischemic heart disease in middle age, are obese, have diabetes, and/or have gout (6). White peri- and postmenopausal women of slender build who avoid dairy products, lead a sedentary life, have a high salt–low protein diet, consume copious quantities of coffee and alcohol, and smoke are at particular risk for osteoporosis (7). The cancer risk factors amenable to modification mainly relate to diet and tobacco, but a multitude of risk factors may be identified that correlate with an increased risk of cancer (8,9). An increased incidence of colorectal carcinoma is found amongst persons over 40 years of age who have a personal or family history of rectal polyps, a personal history of ulcerative colitis, a high fat diet, a low fiber diet (in colonic carcinoma) or a high intake of alcohol (in rectal carcinoma), and/or blood in the feces. Colorectal cancer is the second most common form of cancer and the second most common cause of cancer deaths in the United States. Persons at particular risk for lung cancer are smokers who started smoking at a young age (15 years), who are heavy smokers, who inhale and who have been smoking for many years (notably the over 50-year-old age group), or who have been exposed to asbestos. In 1986, lung cancer was the most common cancer in American men; its prevalence is increasing in women. It is believed that eight in ten cases are attributable to smoking. Factors that are believed to predispose to breast cancer in women include nulliparity or the first child after the age of 30 years, breast cancer in the immediate family, prolonged exposure to hormonal stimuli (early menarche, late menopause), a personal history of breast cancer, a lump or nipple discharge, and possibly a high fat diet. An increased risk of cervical carcinoma correlates with an early commencement of sexual intercourse, multiple partners, uncircumcised male partners, a sexually transmitted disorder (notably herpes genitalis), multiple pregnancies, a lower socioeconomic background, and unusual vaginal bleeding or discharge. Skin cancers are associated with a fair skin, excessive exposure to sunlight, and occupational exposure to pitch, coal tar, or creosote.

Based on a knowledge of mortality and morbidity statistics analyzed according to age and sex and an understanding of the etiopathogenesis or natural history of disease, the clinician can sketch a potential disease picture. Further formulation of a working hypothesis requires patient-specific information. This involves a process of data collection, which essentially subjects the individual to a battery of tests. These tests include questions in the history taking, physical examination procedures, and laboratory and radiological investigations. Such data collection is based on the same diagnostic decision making principles regardless of whether the patient is complaining of overt disease, covert disease is detected in a patient with an unrelated complaint, or a risk factor is noted in a client during a periodic health examination.

Logically one would expect that the more tests used to assess an individual, the more accurate would be the definitive health status assessment report. This is not necessarily the case. The use of numerous laboratory tests, instead of enhancing clinical accuracy, may under certain circumstances paradoxically promote inaccuracy. This can become a particular problem in the specific periodic health examination, the primary practitioner's hypothetico-deductive version of multiphasic community screening. Certain assumptions are made in multiphasic laboratory screening that are not necessarily justified, for example, the notion that the laboratory values of the test follow the bell-shaped gaussian distribution, whereas most biological variables do not conform; the notion that established limits for each test include 95% of the healthy population, whereas limits for laboratory tests are seldom established using appropriately distributed populations; and the assumption that the client tested belongs to the same population to which the frequency distribution applies (10). If gaussian and percentile definitions of normal are applied, then persons whose test results fall outside the range of designated "normal" values are abnormal—possibly diseased? The consequence of this is an upper limit syndrome of nondisease, which in practice translates into the postulate that the only "normal" persons are those who have not undergone sufficient investigations. Given that the "normal" range of a test includes the lower 95% of test results, then the likelihood that an individual subjected to the test will be labeled "normal" is 95% or 0.95; conversely, there is a 5% chance that an abnormal test result will be recorded. If this individual then undergoes two independent tests, the likelihood of being found "normal" is 0.95 × 0.95, or 0.90. Bearing in mind that clinically "normal" test findings are frequently taken to imply a healthy or at least nondiseased state, it may be somewhat disturbing to realize that the likelihood of being labeled "normal" is 0.95 to the power of the number of independent tests performed. A patient subjected to 20 independent tests has a one in three chance of being labeled "normal"; a client undergoing 100 independent screening tests has approximately a six in 1,000 chance of being considered free of "disease."

"Given these mathematics then, the number of false positive results climbs astronomically as more and more tests are added for low prevalence disease" (11). This summarizes the dilemma underlying multiphasic screening of healthy population groups. When multiphasic screening produces "abnormal" test results in apparently well persons, uncertainty about the implications of abnormal test results in healthy people and the statistical likelihood of false-positive results confuse the potential usefulness of multiphasic screening as a disease prevention measure (12). Even in disease-targeted rather than random multiphasic screening, the risk of making a false-positive diagnosis becomes enormous when screening for a rare condition. Despite the availability of tests with good sensitivity and specificity, incorrect labeling of healthy people can occur (1). The B_{27} antigen test has a sensitivity of 90% and a specificity of 95%; the prevalence of ankylosing spondylitis, however, is no greater than 1%. Using radiology as the gold standard for diagnosing ankylosing spondylitis, the positive predictive value of B_{27} serology is 15%; the negative predictive value is 99.9%. In so rare a disorder, of every 100 people with a positive B_{27} test result, 85 do not have ankylosing spondylitis. The potential for misdiagnosis is daunting.

To minimize the likelihood of errors in periodic health examination and case finding, limitations are imposed on both the selection of disorders and the persons considered appropriate for early disease diagnosis; to limit errors in definitive diagnosis, the gestalt of clinical variables takes precedence over mathematical computations. Inaccuracy in probabilistic prediction in clinical thinking may be attributable to the selection of an incorrect anchor point (the average probability of a group) or the failure to individualize correctly. "A correct anchor point probability may be obtained either through knowledge of the literature or by extensive clinical experience . . . effective individualization of predictions requires consideration of social and psychologic, as well as biologic, characteristics of the patient" (13). In evaluating abnormal test results, the clinician should be guided by clinical judgment based on history and examination findings rather than on probability theory applied to laboratory test results. This is equally true whether an attempt is being made to diagnose overt disease or to detect "at-risk" persons. Probability diagnosis can be employed to refine but not replace clinical decision making.

The goal of clinical activity is to identify disease at that stage of its natural history when therapy is most effective or easiest to apply. The central dilemma of screening activities is the failure to detect disease in apparently healthy people. This can be as catastrophic as inappropriate labeling of patients with covert disease. In the former, delayed diagnosis may mean irreversible disease; in the latter, inappropriate labeling can lead to implementation of unnecessary treatment. Unindicated intervention has a cost–benefit ratio that approximates infinity. Incorrect labeling of healthy people can result in unnecessary adoption of the sick role; in diseased persons it may mean inadvertent allocation of the patient to a chronic, rather than temporary, sick role. The sick role has implications both for self-perception and for the obligations and privileges of that individual within society. Inappropriate labeling significantly burdens individuals and society.

The risk of inappropriate labeling can be somewhat reduced by applying stringent criteria to data collection and interpretation when performing the periodic health examination. The periodic health examination, which has largely replaced the more wasteful annual check-up, is widely accepted in primary care. It advocates age- and sex-specific disease screening schedules. The criteria used to determine which conditions should be included in periodic health examination schedules have been well described. The Canadian Task Force, in its investigation of the appropriate use of the periodic health examination in clinical practice, has provided a description of the criteria for the assessment of potentially preventable conditions and produced a classification of recommendations on inclusion or exclusion of these conditions in the periodic health examination (3). In addition to recommending replacement of the annual check-up with an age- and sex-relevant periodic health examination schedule, the Task Force targeted the need for close scrutiny of the effectiveness of proposed interventions. In deciding whether a condition should be judged eligible for inclusion in a periodic health examination schedule, three aspects were considered: the current burden of mortality and morbidity caused by the condition, the validity and acceptability of both the maneuver used to identify the risk of the condition and the maneuver used to effect its prevention; and the effectiveness of the resulting intervention. Evaluation of the proposed detection maneuver should be based on three sets of criteria: the risks and benefits; the sensitivity, specificity, and predictive value of the test; and the safety, simplicity, cost, and acceptability to the patient of the proposed intervention. The effectiveness of intervention was graded according to the quality of evidence obtained:

I. Evidence obtained on at least one properly randomized trial.
II-1. Evidence obtained from well-designed cohort or case-control analytical studies, preferably from more than one research center.
II-2. Evidence obtained from comparison between times or places, with or without intervention; dramatic results on uncontrolled experiments are included in this category.

III. Opinions of respected authorities based on clinical experience, descriptive studies, or reports of expert committees.

The health protection package proposed by the Canadian Task Force for persons aged 16–44 years includes:

- Immunization boosters against polio at 16 years of age, against tetanus and diphtheria (optional) at ten yearly intervals, and against rubella for women intending to bear children but not contemplating pregnancy for the next three months.
- Immunization prophylaxis as indicated for overseas travel.
- Screening of persons at risk for tuberculosis, with immunization as indicated.
- Screening for sexually transmitted diseases; smears for gonorrhea and serology for syphilis in pregnant women and persons with multiple sexual partners.
- Cancer screening at appropriate intervals in people at risk for skin, cervical, and bladder cancer.
- Screening for hereditary conditions such as Tay-Sachs disease, thalassemia, and muscular dystrophy.
- Screening for iron deficiency anemia and malnutrition in women of lower socioeconomic groups and in food faddists.
- Screening for hypertension at five-yearly intervals or more frequently.
- Annual screening for dental caries, periodontal disease, and oral cancer.
- Counseling for family dysfunction and marital and sexual problems.
- Screening and counseling to minimize the risks of alcoholism, smoking, and motor vehicle accidents.

Additional health screening for persons aged 46–64 years includes:

- Annual occult blood tests for colorectal cancer.
- Counseling for retirement distress.

In women of this age group additional screening is recommended for:

- Breast cancer and hypothyroidism.

For persons aged 65–74 years, in addition to more frequent breast cancer, hypertension, and malnutrition screening, the health protection package may also include:

- Annual influenza immunization.

An alternative approach to implementation of a screening package is to consider the following screening frequency recommendations (2–3, 14–17).

At ten-yearly intervals clients should be assessed for:

- Tobacco use. This is particularly important before the age of 45 years.

At four-yearly intervals case finding in adults should include screening for:

- Elevated serum cholesterol. Assessment of total serum cholesterol level

suffices until 50 years of age; thereafter, HDL-cholesterol determinations are also suggested.
- Obesity. The body mass index may be determined, or the actual and ideal body weight compared.
- Hypertension. Elevation of blood pressure above 140/90 mm Hg in three or more serial readings is regarded as diagnostic of hypertension in healthy adults aged 25–40 years.

At two-yearly intervals adult patients should be checked for:

- Hypertension (this refers to healthy adults aged 40–60 years).
- Breast cancer. Professional breast examination of women aged 20–50 years; mammography for high-risk patients under the age of 50 years.
- Cervical cancer. Papanicolaou (Pap) smears should be performed every two years in sexually active women until the age of 70; the second smear, however, should be performed 12 months after the first.

Annual evaluation is recommended to exclude:

- Breast cancer. Professional breast examination in high-risk women under the age of 50 years; professional breast examination and mammography for all women over the age of 50.
- Coronary artery disease. Blood pressure and serum cholesterol levels should be checked in high-risk patients.
- Hypertension. Blood pressure checks are proposed for patients over the age of 60 years.

Self-screening based on practitioner education of healthy persons is suggested for:

- All women. Women should perform breast self-examination at monthly intervals and should check for any abnormal gynecological bleeding premenopausally or any vaginal blood loss postmenopausally.
- All men. Men should palpate for testicular lumps at monthly intervals.
- Oral lesions. All persons discovering any oral lesions that persist for more than one week should seek professional advice.
- Skin lesions. A new or changing skin lesion should be reported.

It is not recommended that healthy people be screened for diabetes, glaucoma, osteoporosis, chronic obstructive airway disease, thyroid dysfunction, tuberculosis, sexually transmitted diseases, and cancer of the lung, bladder, ovaries, or prostate. While some would argue that the value of various listed procedures currently lacks clinical justification, others regard the application of such inclusion criteria as excessively stringent.

Authoritative bodies such as the American Cancer Society propose a less restrictive approach. The American Cancer Society recommends the following routine screening protocol (18):

- Pap smears at least every three years until age 65.
- A pelvic examination every year for women over 40.

- A pelvic examination and Pap smear in all women at menopause; endometrial sampling for those in high-risk groups.
- Annual mammography for women over 50.
- Annual professional breast examination for women over 40.
- Monthly breast self-examination for all women.
- Annual stool guaiac slide tests for everyone over 50.
- Sigmoidoscopy every three to five years for everyone over 50, and then annually after two examinations with normal findings.
- Annual digital rectal examination for everyone over 40.
- Annual check-ups for everyone over 40, including cancer counseling and examination of the thyroid, testicles, prostate, lymph nodes, oral region, and skin.

This more comprehensive approach of the American Cancer Society sacrifices the overall cancer screening efficiency rate in order to detect a few cancers that would be missed when the more stringent criteria of the Canadian Task Force are applied. Implementation of the Cancer Society's protocol implies an increased cost to the general community with increased benefit for a few. In view of the discrepancies between the recommendations of two such authoritative bodies, it may be deemed prudent to seek clarification by consulting Frame's (16) most recent comment.

Cognizance of both costs and benefits forms an integral part of the approach proposed by Frame and Carlson (2) to justify active case finding and adult health maintenance in primary practice. Although it is not always practical to meet all the theoretical prerequisites, this approach does provide a rational guideline in clinical practice. Before case finding and health maintenance are justified, Frame and Carlson propose that the following criteria be met (2):

- The disease has a significant effect on the quality and quantity of life.
- Acceptable methods of treatment are available.
- The natural history of the disease incorporates an asymptomatic period during which intervention significantly reduces mortality or morbidity.
- Intervention during the asymptomatic period yields a result superior to that obtained by delaying treatment until the onset of symptoms.
- Tests allowing recognition of the asymptomatic condition are available at reasonable cost.
- The incidence of the condition is sufficient to justify the cost of case finding.

An analysis of a condition according to these criteria ideally requires an awareness of several other factors.

1. A knowledge of the natural history of the condition. More specifically, adequate decision making requires data pertinent to the incidence and prevalence of the disease; an understanding of the risk factors associated with the etiology and pathogenesis of the disorder; and a knowledge of the prognosis of the untreated condition. In the case of cancer of the uterine cervix, it is necessary to differentiate between invasive carcinoma and cancer-in-situ (19). Since the introduction of the Pap smear the incidence of

carcinoma-in-situ, now a clinically recognizable condition, is three times that of invasive cervical cancer. The incidence of carcinoma-in-situ peaks at 34 years of age and then falls rapidly after the age of 40; the prevalence of invasive cervical cancer rises to 28 in 100,000 women after the age of 40 years. It is generally agreed that a proportion of carcinomas-in-situ progress to invasive carcinoma; the time taken for this progression is the subject of much controversy. Estimates range from 5.8 to 10 years. Risk factors for cancer of the cervix include frequent sexual intercourse, multiple sexual partners, first intercourse at a young age, lower socioeconomic status, and being a black woman.

2. Objective data on the outcome of therapeutic intervention at various stages of the disease process, i.e., information about the potential for early intervention to positively influence disease prognosis. Carcinoma-in-situ has a five-year survival rate of almost 100%. Up to 60% of carcinomas-in-situ may not progress to invasive cancer; those that do have a significantly poorer survival rate. Stage I invasive cancer has a five-year survival rate of 80%; the five-year survival rate for women with stage III cancer is 30%.

3. An assessment of the costs of treating symptomatic disease compared with the costs of screening an asymptomatic population. The Pap smear has a reasonable financial cost and is acceptable to both clients and clinicians. The only real controversy surrounding its use is the frequency with which it should be performed. The variables that affect this decision are the sensitivity of the test and the rate of disease progression. Frame's current recommendation is screening every two years in sexually active women after they have been screened annually for two years. Screening should continue until the age of 70. Interestingly, Frame recommends more frequent screening than either the Canadian Task Force (screening every five years in women aged 35–60 years) or the American Cancer Society (screening at least every three years).

Similar discrepancies are encountered when Frame applies his criteria to a number of prevalent conditions. Lung cancer deaths are escalating, yet neither Frame nor the Canadian Task Force recommends screening for this neoplasm. Nor do they recommend screening for prostate cancer in asymptomatic men over the age of 40 years, despite prostate cancer having the second highest incidence of all cancers in men. Although the efficiency of digital rectal examination in asymptomatic clients is highly questionable, rectal examination remains a better screening test than either acid phosphatase determinations or cytology in males with evidence of urinary tract obstruction. Digital rectal examination for the early detection of colorectal carcinoma enjoys less favor; virtually untested as a screening method for colorectal carcinoma, this examination may miss up to 90% of neoplasms. Sigmoidoscopy is effective for detecting early lesions but is expensive and traumatic for clients; up to 40% of colon cancers may be missed. Stool testing for occult blood is acceptable to clients and is convenient; effectiveness is inconclusive. Frame recommends that persons between the ages of 40 and 50 should have a biannual six-slide stool occult blood test, and the test should be conducted annually thereafter (19); the Canadian Task Force proposes annual stool occult blood testing after the age of 46 years.

Frame's proposals are not always so congruent with the recommendations of the Canadian Task Force. With regard to coronary artery disease prevention, Frame recommends that risk factors such as tobacco use, elevated serum cholesterol, and hypertension be sought and modified (20). The Canadian Task Force does not recommend screening for tobacco use and hypercholesterolemia but does suggest that blood pressure be assessed on each visit to a clinician and not only at periodic health examinations. For adult health maintenance, Frame proposes blood pressure assessment at two-yearly intervals. The detection and treatment of hypertension is also recognized as the best method for preventing stroke.

Frame's recommendations are not always more liberal than those of the Canadian Task Force. He recommends annual vaccination for all persons at high risk for lower respiratory tract infection (21). The Canadian Task Force would prefer to extend influenza immunization to anyone over the age of 64 years. Frame's recommendations are in some aspects more comprehensive than those of the Canadian Task Force, which fails to review osteoporosis. Osteoporosis is perhaps best addressed by case finding in high-risk groups. While Frame is opposed to periodic health screening for osteoporosis, he supports clinical evaluation of menopausal women for osteoporosis risk (22).

The application of probabilistic reasoning to disease prevention and health promotion provides a framework wherein the rationale underlying various procedures can be analyzed. While not providing irrefutable answers, it does provide a standard against which progress can be measured. Fallacies in clinical thinking can be exposed by implementing the use of measurable criteria. Only criterion-referenced probabilistic reasoning permits an objective demonstration of the fallacy that early diagnosis always "improves" survival, even when therapy is ineffective. Cancer survival rates, measured over a five-year time span, are improved regardless of therapy when the cancer is diagnosed early. Earlier diagnosis merely means earlier identification of the disease and provides a longer time span in which to clinically scrutinize its uninterrupted natural history. Without adequate recording of pertinent data and analytical scrutiny and without controlled randomized trials, lead-time bias and preferential detection of slow-growing tumors (length-time bias) can result in fallacious clinical thinking. Early diagnosis is not inherently a justifiable procedure; early diagnosis of a condition that lacks a feasible intervention merely creates anxiety. Probabilistic reasoning helps to identify which conditions justify diagnosis during the asymptomatic phase of their natural history; it also facilitates the accurate diagnosis of clinically overt disease.

The risk of inappropriate labeling can be somewhat reduced by maintaining a clear perspective about the clinical implications of the data collected. While probabilities have been more clearly calculated for the outcomes of special investigations, the most important clinical data, and sometimes the only necessary data, are derived from history taking and physical examination. The same principles and probability strategies apply equally to information derived from history taking, physical examination, laboratory investigation, and radiological evaluation. Pattern recognition suggests a number of possibilities, and hypothetico-deductive reasoning refines these diagnostic postulates by eliminating the least likely conditions. Each piece of clinical information collected has its own inherent potential to support retention or elimination of the working hypothesis. A sensitive test gives positive results when the disease is truly present; it seldom has false negatives. A specific test is one that has negative results when the disease is absent; it seldom has false positives. Sensitive tests are important when apparently healthy people are screened for disease; specific tests are used to confirm selection of a diagnostic label before therapy is initiated. The sensitivity and specificity of clinical tests can be calculated by measuring test results in a group of persons in whom the disease is already proven to be either present or absent. While sensitivity and specificity information relates to evaluation of a diagnostic test, posttest probability is more useful to diagnostic decision making. Using the same data required for ascertaining test sensitivity and specificity, the posttest probability of disease following a positive test result and the posttest probability of no disease following a negative test result can be calculated.

A 70-year-old man presents to his chiropractor complaining of low back pain. On history taking he is found to have nocturia, a poor urinary stream, and urinary hesitancy. The chiropractor decides to perform a rectal examination, and on palpation finds a nodular hard prostate. The sensitivity of rectal examination for prostatic carcinoma is rated as 69%; the specificity of this test is 89% (23). In a population in which the prevalence of prostatic carcinoma is 50%, the posttest probability of this patient having prostatic carcinoma is increased to 86%. A positive result from rectal examination increases the pretest probability by 36%. If prostatic examination had found an enlarged prostate with a median sulcus and firm lateral lobes, then the test would have been negative for prostatic carcinoma, but probability reasoning shows that the posttest probability of prostatic carcinoma remains high (74%) and further investigations are required to support either this or some other working hypothesis (Fig. 2.1). Compared with rectal examination, the acid phosphatase test has a sensitivity of 56% and a specificity of 94%. A raised acid phosphatase in the patient presenting with low back pain and prostatism has a posttest predictive value for prostate cancer of 90%; a normal acid phosphatase level has a negative posttest predictive value of 68%. In this patient, the probability of prostate cancer being present is increased by 40% following the finding of a raised acid phosphatase; the likelihood of prostate cancer is increased by 18% following a negative test result (Fig. 2.2). In practice, clinical reasoning seeks to confirm rather than refute the working hypothesis. Additional test results are therefore sought in order to support a working diagnosis; only when tests fail to support a working diagnosis is it discarded and another working diagnosis postulated. Tests are then performed in order to seek clinical support for this new postulate before arriving at a definitive diagnosis. This clinical approach is contrary to the hypothetico-deductive approach used in the basic sciences, where experiments are performed in order to disprove the working hypothesis.

Most is gained from any single clinical sign, symptom, or laboratory or radiological test when the prevalence or pretest probability is between 40% and 60%. If prostatic examination is used in a specialty center for prostatic diseases of elderly males where the prevalence of prostate cancer is 91%, then the predictive probability of a positive result on rectal examination is in-

Rectal examination: Test sensitivity = 69% = $a/(a+c)$
Test specificity = 89% = $d/(d+b)$
Positive predictive value (posttest likelihood of a positive test result in patients with the disease) = $a/(a+b)$
Negative predictive value (posttest likelihood of a negative test result in clients without the target disease) = $d/(c+d)$
Probability that a patient does have the disease despite a negative test result = $c/(c+d)$ or $(1-d)/(c+d)$

PRETEST PREVALENCE 50%
Positive predictive value = 86%
Negative predictive value = 74%
Probability that a patient has the target disease despite a negative test result = 26%

		PROSTATE CANCER*	
		Present	Absent
Rectal Examination	Positive	69 (a)	11 (b)
	Negative	31 (c)	89 (d)

PRETEST PREVALENCE 91%
Positive predictive value = 98.4%
Negative predictive value = 22%
Probability that a patient has the target disease despite a negative test result = 78%

		PROSTATE CANCER*	
		Present	Absent
Rectal Examination	Positive	690 (a)	11 (b)
	Negative	310 (c)	89 (d)

PRETEST PREVALENCE 2.4%
Positive predictive value = 13.6%
Negative predictive value = 99.1%

		PROSTATE CANCER*	
		Present	Absent
Rectal Examination	Positive	69 (a)	440 (b)
	Negative	31 (c)	3,560 (d)

*Note: The gold standard for diagnosis is rectal biopsy.

Figure 2.1. Posttest predictability of prostate cancer using rectal examination.

Acid phosphatase: Test sensitivity = 56% = $a/(a+c)$
Test specificity = 94% = $d/(b+d)$
Positive predictive value (posttest likelihood of a positive test result in patients with the disease) = $a/(a+b)$
Negative predictive value (posttest likelihood of a negative test result in clients without the target disease) = $d/(c+d)$
Probability that a patient does have the disease despite a negative test result = $c/(c+d)$ or $(1-d)/(c+d)$

PRETEST PREVALENCE 50%
Positive predictive value = 90%
Negative predictive value = 68%
Probability that a patient has the target disease despite a negative test result = 26%

		PROSTATE CANCER*	
		Present	Absent
Acid Phosphatase	Positive	56	6
		a	b
		c	d
	Negative	44	94

PRETEST PREVALENCE 91%
Positive predictive value = 98.9%
Negative predictive value = 17.5%
Probability that a patient has the target disease despite a negative test result = 82.5%

		PROSTATE CANCER*	
		Present	Absent
Acid Phosphatase	Positive	560	6
		a	b
		c	d
	Negative	440	94

PRETEST PREVALENCE 2.4%
Positive predictive value = 19%
Negative predictive value = 99%

		PROSTATE CANCER*	
		Present	Absent
Acid Phosphatase	Positive	14	60
		a	b
		c	d
	Negative	11	940

*Note: The gold standard for diagnosis is rectal biopsy.

Figure 2.2. Posttest predictability of prostate cancer using acid phosphatase.

creased to 98.4%. While this patient is highly likely to have prostate cancer, a positive finding only increases the pretest probability of disease by 7.4%. The likelihood of prostate cancer in the presence of normal findings on rectal examination is 78%; a negative test only reduces the pretest probability by 13% (Fig. 2.1). In this population, the positive predictive value of acid phosphatase is 98.9%, the negative predictive value 17.5% (Fig. 2.2). A positive test result only marginally increases the likelihood of disease in a high-risk group; a negative result fails to exclude prostate cancer from the differential diagnosis. A negative test result fails to provide an explanation for the clinical presentation of the patient. The relatively high specificity of these tests, particularly if used sequentially, does reduce the risk of erroneously detecting prostate cancer in a benign prostate.

Conversely, rectal examination may be included as a case finding procedure in an asymptomatic population with a low prevalence for prostatic carcinoma (2.4%). Under these circumstances, rectal examination has a positive posttest predictability of 13.6% and a negative posttest predictability of 99.1%. The likelihood of prostate cancer in an individual with a negative test result is 1%. In this asymptomatic population group, an individual found to have a normal prostate on rectal examination is unlikely to have prostate cancer; but rectal examination only reduces the probability that this condition is present by 1.4% (Fig. 2.1). Acid phosphatase has a positive predictive value of 19% and a negative posttest predictive value of 99% in this population (Fig. 2.2). As prevalence falls, the positive predictive value of a test decreases and the negative predictive value increases. The limitations inherent in the use of these tests as screening procedures are further compounded by their somewhat indifferent sensitivity rates. When screening an asymptomatic population, tests with a high sensitivity are required so as to adequately detect subclinical disease states.

In attempting to label a condition in a patient who seeks consultation for a specific complaint, the likelihood ratio has also been found to provide a useful, and increasingly popular, means of probabilistic thinking. Likelihood ratios, like positive predictive values, provide information about the possibility that a positive test result and target disease are simultaneously present. Unlike the positive predictive value, likelihood ratios are not influenced by disease prevalence. Likelihood ratios can also take into account the strength of the original mode of data collection. Nominal data provide information on what is essentially a binary scale; the prostate is or is not enlarged. Ordinal data attempt to quantify the hardness of the prostate: firm, hard, stony hard. Interval data, as derived from laboratory tests, add additional precision to probability analysis of clinical data. These data provide a variety of cutoff points against which the probability of disease can be computed. The likelihood ratio expresses the odds that a given level of a diagnostic test result could be expected in a patient with a target disorder (1).

The likelihood ratio of positive rectal examination findings in our patient presenting to the chiropractic clinic is 6.27 (Fig. 2.3). The likelihood ratio of a hard nodular prostate found on rectal examination being prostate cancer is the ratio of the true-positive to the false-positive results. This is the same as calculating the ratio between the sensitivity of the test and one minus its specificity. The likelihood ratio provides information about the diagnostic

Likelihood Ratio:
The likelihood ratio of prostate cancer following a positive rectal examination finding:

$$\frac{\text{True-positive rate}}{\text{False-positive rate}} = \frac{\text{Sensitivity}}{1 - \text{Specificity}} = \frac{0.69}{0.11} = 6.27$$

The likelihood ratio of prostate cancer following a positive acid phosphatase test result:

$$\frac{\text{True-positive rate}}{\text{False-positive rate}} = \frac{\text{Sensitivity}}{1 - \text{Specificity}} = \frac{0.56}{0.06} = 9.3$$

Posttest Odds:
Posttest odds for target disorder = pretest odds (for target disorder) × likelihood ratio (diagnostic test result)
Posttest odds in a patient with a:

- 50:50 chance of prostate cancer before finding a craggy prostate on rectal examination is 1 × 6.27 = 6.27.
- 86 in 100 chance of prostate cancer before finding a positive acid phosphatase laboratory result is 6.27 × 9.3 = 58.3.

Posttest Probability:
Probability expressed as odds:

$$\text{Pretest odds} = \frac{\text{Pretest probability}}{1 - \text{Pretest probability}} = \frac{0.5}{0.5} = 1$$

$$\text{Posttest probability} = \frac{\text{Posttest odds}}{\text{Posttest odds} + 1} = \frac{58.3}{59.3} = 0.98$$

After sequential testing, this patient's posttest probability of prostate cancer is over 98%; the likelihood that this patient's diagnosis is prostate cancer is substantially increased from 50% to 98% following sequential testing.

Figure 2.3. Sequential testing using likelihood ratios and posttest probabilities.

test; once again the clinician desires information about the likelihood of a target disease in a particular individual. The likelihood ratio can be converted to the posttest odds for a target disorder. To calculate this, the pretest odds for prostate cancer are multiplied by the likelihood ratio of the positive test result in the clinic patient. The pretest odds are the pretest probability or prevalence of the disease divided by one minus the pretest probability; the posttest probability is determined by dividing the posttest odds by the sum of one plus the posttest odds. This is 86% in our clinic patient. Applying a similar strategy to the acid phosphatase test, the likelihood ratio of a positive test result in a patient with prostatic carcinoma is 9.3. The posttest probability of prostate cancer in our clinic patient is 90%. Before this patient had his rectal examination, his pretest probability of prostate cancer was one; after the clinician found a craggy, nodular prostate on rectal examination, the patient's posttest odds rose to 6.27. With sequential testing, the pretest odds of prostatic cancer are now 6.3. After a positive acid phosphatase test result, the posttest odds of prostate cancer reach 58.3. This patient's posttest probability of prostate cancer after positive findings on rectal examination and the serum acid phosphatase test is over 98%. By moving from less invasive to more hazardous and costly procedures, the diagnosis of prostate cancer becomes increasingly probable and can be finally confirmed by transrectal biopsy or therapeutic surgery. Diagnostic labeling is the precursor to intervention, and thus it is vital that the correct label is applied. As diagnostic procedures themselves have a cost–benefit ratio, selective use must be made of these procedures. Although transrectal biopsy is a single step to a definitive diagnosis, it is an invasive procedure and only indicated when other less traumatic procedures suggest that it is clinically justified. Probability reasoning helps to quantify diagnostic decision making.

Probabilistic reasoning not only provides a basis for diagnostic decision making, but also is the foundation of rational therapy. The essence of selecting appropriate clinical intervention is to achieve a therapeutic outcome in which the benefits outweigh the costs. Benefits are determined by comparing actual clinical progress with the expected natural progression of the disorder. Costs are determined by ascertaining whether the known side effects of therapy are causing adverse side effects in any particular patient.

Before concluding that an intervention is effective, it should be demonstrated that the prognosis of the condition is changed and that this change is attributable to the implemented intervention strategy. Rarely does an intervention so dramatically alter outcome that all parties are unanimously enthusiastic about its effectiveness. Any perceptible responses must therefore be diligently recorded. Having defined the criteria that are to be considered indicative of improvement, one must determine whether such changes are attributable to the proposed intervention. The possibility that an intervention is effective becomes increasingly probable when:

- The intervention always precedes the designated improvement response.
- There is a direct quantifiable intervention–response relationship.
- The outcome has a biologically plausible explanation in terms of the intervention offered; this may be further supported by animal experimentation.
- The findings of this intervention–outcome relationship are consistent with information available from other epidemiological and clinical trials. When similar results have been obtained from randomized clinical trials, there is a strong probability that the intervention is effective; support from cohort studies, and to a lesser extent from case-control studies, also significantly increases the possibility that an effective intervention is being implemented. Case studies create interesting ideas that only further research can dignify with possibility and probability quotients.
- There is a strong association between the outcome and the intervention. Randomized trials and cohort studies permit calculation of the relative

strength of interventions by comparing outcome rates in treated and untreated patients. The odds ratio, which quantifies the probability that an intervention is effective, is calculated by comparing the proportion of test patients who improve with intervention with the proportion of patients who improve without intervention (Fig. 2.4).

The less profound the clinical impact of an intervention, the greater the necessity for careful analysis of the outcome using accepted epidemiological and clinical trial procedures. Because the outcome of an intervention is often equivocal, statistical significance is often used to ascertain whether any recorded response could be attributed to chance. Such statistical analysis is not immune to error. Type I errors occur when the clinical researcher incorrectly concludes that a therapy achieves a benefit greater than that which can be attributed to chance; type II errors exemplify the erroneous rejection of an efficacious therapy. Type I errors may be manifestations of the placebo effect; type II errors exemplify the "tomato effect" (24,25).

An effective intervention may be defined as a clinical management strategy that provides the patient with a greater benefit than that which can be derived from chance alone. Before the efficacy of an intervention can be determined, the characteristics of efficacious intervention must be operationally defined. Adequate operationalization of the criteria whereby efficacy

---

A Randomized Trial

LOW BACK PAIN

	Improvement	Unchanged
Adjustment	a	b
No adjustment	c	d

Relative probability that the adjustment administered is effective in the management of low back pain in this group of patients is:

$$\frac{a}{a+b} \div \frac{c}{c+d}$$

In order to determine the effectiveness of chiropractic adjustment in this group of patients with low back pain, the proportion of patients that improved following manual adjustment is compared with the proportion that improved without chiropractic clinical intervention. It is only the clarity with which criteria including and excluding individuals are stated, and the stringency with which these criteria are implemented in selecting subjects, that enables extrapolation of these findings to other nontrial patients suffering from low back pain. Only patients with similar characteristics of and prognoses for low back disease have the calculated relative probability of experiencing improvement.

Figure 2.4. Assessing the effectiveness of intervention.

can be ascertained will help to clarify the confusion surrounding biological and statistical significance. The measurement of "benefit" is an entirely clinical matter; the probability that any clinically defined benefit is likely to occur involves computation. A statistically significant response should not be confused with a biologically trivial outcome; nor is a statistically insignificant response necessarily biologically unimportant. Just as diagnostic tests are ideally both sensitive and specific, so should interventions achieve biological and statistical significance.

When determining the utility of an intervention both costs and benefits are assessed. Assessing the risks and costs associated with various interventions can be performed along similar lines to those used in assessing benefits. While the safety of clinical interventions is often glibly referred to, its assessment is an extensive task. Adverse reactions, when applied to pharmaceutical therapy, include an array of undesirable reactions variously named: cumulations, tachyphylaxis, tolerance, intolerance, toxicity, overdosage, side effects, secondary effects, idiosyncrasy, hypersensitivity, and allergic reactions (26). Many of the clinically unacceptable side effects associated with chiropractic adjustments are listed amongst the contraindications to manipulation (27).

In deciding whether an intervention should be implemented in a particular patient, one should evaluate any documentation of adverse effects associated with the use of the proposed intervention, and alternative therapeutic choices that may be used in lieu of the proposed intervention. Once an intervention has been implemented, if an untoward alteration in the patient's condition is noted, it is important to ascertain whether the intervention is having an adverse or a beneficial effect.

1. Adverse effect. The question of cause must be addressed. The likelihood that the intervention is responsible for the deterioration in the patient's condition is increased if improvement follows cessation of the intervention and reproduction of the unwanted side effect occurs on reintroduction of the intervention. The viability of a strategy that reintroduces a suspect therapy is determined by the implications of the untoward effect and the range of therapeutic options available for treating the underlying condition.

2. Beneficial effect. Any improvements in the patient's condition should be evaluated in terms of biological plausibility and the time lapse between therapy and improvement. In the final analysis, worsening of the patient's condition on cessation of the therapy, followed by improvement on its reintroduction, greatly increases the odds that the intervention is effective. An understanding of the natural history of the condition is basic to any such approach. This clinical strategy is used with good effect in scientific medicine; it is the basis of the cross-over double-blind placebo-controlled clinical trial.

Probabilistic reasoning has contributed to:

- The identification of variables associated with diverse conditions. It has also been helpful in providing insight into the strength of such associations.
- The rational selection of diagnostic tests in clinical practice. Test selection based on sensitivity and specificity can refine clinical evaluation, limiting unnecessary and costly investigations.

- The quantification of diagnostic decision making by providing positive predictive values and likelihood odds that the patient under investigation has a particular disease or condition.
- The objective evaluation of diverse interventions in clinical care. Probabilistic reasoning may be equally helpful in evaluating therapeutic benefit, safety, and side effects.

Probabilistic reasoning is a basic clinical decision making strategy that contributes to an optimal clinical outcome. Clinical outcome is itself determined by:

- The natural history of the disease.
- The social support system available to the sick patient.
- The psychological makeup and coping skills of the individual.
- The stage at which the condition is diagnosed.
- The accuracy of the definitive diagnosis.
- The availability and efficacy of interventions that may be used to treat the diagnosed condition.
- The clinical performance of the clinician; this includes a consideration of the gestalt of patient–practitioner interaction and the placebo effect.
- The compliance of the patient with the proposed intervention and other advice offered by the clinician.

Probabilistic reasoning pervades clinical practice.

References

1. Sackett DL, Haynes RB, Tugwell P: *Clinical Epidemiology. A Basic Science for Clinical Medicine.* Boston, Little, Brown & Co, 1985.
2. Frame PS, Carlson SJ: A critical review of periodic health screening using specific screening criteria. *J Fam Pract* 2:29–34, 1975.
3. Canadian Task Force. The periodic health examination. *Can Med Assoc J* 121:1193–1254, 1979.
4. National Center for Health Statistics. *Monthly Vital Statistics Report.* Washington, DC, U.S. Government Printing Office, 1988, vol. 87, no. 1.
5. Christie D, Gordon I, Heller R: *Epidemiology.* Kensington, NSW, New South Wales University Press, 1987, pp 8–11.
6. Hall E, Magnus P: Preventive strategies for heart disease. *Aust Med* 1:8–9, 1989.
7. McDonnell JM, Lane JM, Zimmerman PA: Osteoporosis: definition, risk factors, etiology and diagnosis, *AAOHN J* 35:527–530, 1987.
8. Doll R, Peto R: The causes of cancer: quantitative estimates of avoidable risks of cancer in the United States today. *J Natl Cancer Inst* 66:1191–1308, 1981.
9. Jamison JR: *The Client's Handbook of Health Promotion.* Melbourne, PIT Press, 1984.
10. Phillips WS, Thompson DJ: Multi-test screening and the chances of being normal. *J Fam Pract* 11:1095–1098, 1980.
11. Thompson RS: Approaches to prevention in an HMO setting. *J Fam Pract* 9:71–80, 1979.
12. Vogt TM: Risk assessment and health hazard appraisal. *Annu Rev Public Health* 2:31–47, 1981.
13. Shapiro AR: The evaluation of clinical predictions. *N Engl J Med* 298:1509–1514, 1977.
14. Madlon-Kay DJ: Improving the periodic health examination: using a screening flow chart for patients and physicians. *J Fam Pract* 25:470–473, 1987.
15. Canadian Task Force. The periodic health examination 2:1987 update. *Can Med Assoc J* 138:618–626, 1979.
16. Frame PS, Lawrence RS, Taplin S: What's new in health screening. *Patient Care* 21:109–150, 1987.

17. Lindberg SC: Adult preventive health screening: 1987 update. *Nurse Pract* 12:19–29, 1987.
18. American Cancer Society. *Guidelines for Cancer-related Checkups: Recommendations and Rationale.* Chicago, American Cancer Society, 1980.
19. Frame PS: A critical review of adult health maintenance: Part 3, Prevention of cancer. *J Fam Pract* 22:511–520, 1986.
20. Frame PS: A critical review of adult health maintenance: Part 1. Prevention of atherosclerotic diseases. *J Fam Pract* 22:341–346, 1986.
21. Frame PS: A critical review of adult health maintenance: Part 2. Prevention of infectious diseases. *J Fam Pract* 22:417–422, 1986.
22. Frame PS: A critical review of adult health maintenance: Part 4. Prevention of metabolic, behavioral, and miscellaneous conditions. *J Fam Pract* 23:29–39, 1986.
23. Guinan P, Bush I, Ray V, et al: The accuracy of rectal examination in the diagnosis of rectal carcinoma. *N Engl J Med* 2:321, 1981.
24. Podell RN: The 'tomato effect' in clinical nutrition. *Postgrad Med* 76:49–65, 1984.
25. Goodwin JS, Goodwin JM: The tomato effect. *JAMA* 251:2387–2390, 1984.
26. Feinstein AR: *Clinical Biostatistics.* St. Louis, CV Mosby, 1977, pp 256–270.
27. Kleynhans AM, Terrett AGJ: The prevention of complications from spinal manipulative therapy. In Glasgow EF, et al (eds): *Aspects of Manipulative Therapy.* London, Churchill Livingstone, 1985, pp 161–175.

3
Pattern Recognition: A Signpost to Management

Syndrome diagnosis is based on identifying the presence of a variety of clinical, laboratory, and/or radiological features in the patient. It utilizes pattern analysis in a gestalt approach to clinical decision making. Pattern recognition is the reflex response of the experienced clinician; it is the instantaneous realization that the patient's presentation conforms to a typical nosological disease description. While pattern recognition provides the working hypothesis, it is further hypothetico-deductive analysis that converts the implied possibility of an initial working hypothesis into the probability of a definitive diagnosis. Syndrome diagnosis is a popular strategy in clinical decision making for a patient complaining of joint pain.

Joint pain is a prevalent complaint in modern society; up to four in five persons of retirement age may have some rheumatic complaint (1). Rheumatic diseases do not necessarily involve joints; nonarticular disorders include tendonitis, fibrositis, degenerative disc disease, carpal tunnel syndrome, adhesive capsulitis, shoulder-hand syndrome, polymyalgia rheumatica, and giant cell arteritis (2). Furthermore, complaints of stiff and painful joints do not inevitably indicate arthritis: soft-tissue disorders should be considered. Based on prevalence findings in a Swedish rheumatology clinic, the probability of a patient having a soft-tissue rather than a joint problem is approximately one in five; the likelihood that the soft-tissue problem is a bursitis is over 50% (3). Probability reasoning is a useful guide in focusing diagnostic effort; further diagnostic decision making requires a consideration of the clinical picture presented by the patient. Probability reasoning in differential diagnosis is based on the distribution of disease in population groups; only a search for the presence or absence of syndrome criteria presenting as signs and symptoms in a particular patient can lead to diagnostic labeling of that patient. Exclusion of a condition is as successfully achieved by syndrome or pattern comparison as is diagnosis of a disease. In excluding soft-tissue lesions from the differential diagnosis of a patient with joint pain, it is useful to know that patients with bursitis are likely to:

- Complain of joint pain; on examination a full range of motion is present despite knee, hip, or shoulder pain.
- Admit to pain above and below the joint during physical examination.
- Complain of night pain. Direct pressure in the region of the involved "joint" results in discomfort; this tenderness is of sufficient severity to wake a sleeping person.
- Find that flexion aggravates the lesion.
- Experience exquisite tenderness when the examiner exerts slight pressure on a particular area in the vicinity of the joint. In "knee" pain this trigger point is found on the medial side of the tibia about 2 or 3 inches below the joint margin over the insertion of the pes anserinus. In "hip" pain the tender point lies over the greater trochanter. In "shoulder" pain pressure applied laterally just under the acromion precipitates severe pain.

The absence of these findings largely excludes a diagnosis of bursitis; their presence is suggestive of bursitis. Although the absence of these criteria does narrow the range of possible explanations for the joint pain, their presence cannot exclude arthritis.

Syndrome diagnosis depends initially on the detection of inclusion criteria. Whereas experimental science seeks to disprove a hypothesis, hypothetico-deductive reasoning in clinical practice seeks largely, but not exclusively, to find signs and symptoms that support the working hypothesis. A positive diagnosis of arthritis is based on the presence, rather than the absence, of various criteria. The demonstration of impaired joint motion and tenderness in the region of the joint margin and/or pain on weight bearing are suggestive of an arthritic lesion. Only when various inclusion criteria have been detected may the absence of exclusion criteria be used to further refine the working hypothesis. Once true joint involvement is established, dichotomous anatomical reasoning may be used to allocate the patient to one of two groups. A patient with monoarticular arthritis is suspected of osteoarthritis, septic arthritis, gout, pseudogout, tumor, neuropathic arthropathy, juvenile rheumatoid arthritis, or early rheumatoid arthritis. A patient with polyarticular arthritis may have rheumatoid arthritis, juvenile rheumatoid arthritis, psoriatic arthritis, Reiter's syndrome, systemic lupus erythrematosus, hypertrophic pulmonary osteodystrophy, or gout (2). Such interim reasoning is then supplemented by further diagnostic analysis. In the case of Reiter's syndrome, pattern diagnosis requires that this label only be applied to patients who also have nonspecific urethritis, mucocutaneous lesions, and conjunctivitis or, less commonly, uveitis. Not all disorders, however, are equally amenable; the definitive diagnosis of osteoarthritis is more complex. The categorization used suggests that osteoarthritis is a condition affecting single joints; the clinical reality of osteoarthritis is that more than one joint may be involved.

Anatomical reasoning applied to osteoarthritis is more helpful if the distribution of joint involvement in the patient is compared with the probabilistic anatomical charting of joints involved in this condition. Nosological knowledge suggests that osteoarthritis most commonly affects (4):

- Distal interphalangeal joints; Heberden's nodes may occur in adults of any age.
- Proximal interphalangeal joints; Bouchard's nodes are Heberden's nodes at a different site.
- The first carpometacarpal joint.
- The acromioclavicular joint.
- The lower cervical and/or lower lumbar areas. Involvement of the posterior apophyseal joints and associated intervertebral disc narrowing with resultant narrowing of the disc and joint space may result in local or referred pain. Osteophyte formation may precipitate radicular pain due to nerve root compression, myelopathy following spinal canal stenosis, or even stenosis of vertebral artery foramina with evidence of vertebral artery insufficiency.
- The hip; involvement is usually unilateral.
- The knee; knees are commonly involved.
- The first metatarsophalangeal joint; associated bunion formation is common.

In any of these anatomical areas, clinical symptoms such as mild early morning stiffness, pain that worsens on joint use, stiffness following periods of rest, and loss of function strongly suggest the likelihood of osteoarthritis. Clinical signs of local tenderness, soft-tissue swelling, joint crepitus, bony swelling, and restricted mobility confirmed on x-ray by joint space narrowing, osteophytosis, altered bone contour, bony sclerosis, bone cysts, and/or periarticular calcification confirm the diagnosis (5). Unfortunately, the signs that confirm the diagnosis are only detectable late in the natural history of the disease. Marked discrepancies between radiological findings and symptomatology have made it necessary for statistical data to specify on which diagnostic criteria the stated prevalence and incidence rates are based. The pattern diagnosis of osteoarthritis is rich in clinical variables; it lacks quantification of the criteria required before a diagnostic label is attached.

The prevalence of rheumatoid arthritis also varies according to the criteria required for diagnostic labeling. Depending on the classification used, the prevalence of rheumatoid arthritis in the United States varies between 3.0% and 0.25% of the total adult population (6). The American Rheumatism Association has addressed the dilemma of probability statistics being impaired by imprecise diagnostic labeling. They have developed a scheme whereby the likelihood of rheumatoid arthritis is classified as probable in the presence of three signs, definite when five, and classical when seven signs are present. The 11-point diagnostic list follows (7). All findings marked with an asterisk score only when present continuously for six or more weeks.

- Morning joint stiffness.* This is present in two-thirds of cases.
- Joint tenderness* or pain precipitated by movement* affecting one or more joints.
- Soft-tissue thickening* or effusion* of one (score 1) or more (score 2) joints.

- Symmetrical joint swelling*, except for distal interphalangeal joint swelling.
- Subcutaneous nodules.
- Serology positive for rheumatoid factor.
- Synovial fluid that demonstrates poor mucin clotting.
- Characteristic changes of synovium (score 1) and/or subcutaneous nodule(s) (score 1) on histology.
- Typical radiological changes; x-rays are usually normal in the early stages of the disease. The four characteristic radiological signs of rheumatoid arthritis include:
  - Fusiform soft-tissue swelling attributable to the accumulation of intraarticular synovial inflammatory tissue or fluid with capsular distension and soft-tissue edema.
  - Periarticular osteopenia, an early response to hyperemia and disuse atrophy.
  - Diffuse loss of interosseous space; radiological evidence of prolonged inflammation.
  - Marginal erosion of bone; this correlates anatomically with areas where bone is no longer covered with articular cartilage.

More specific classifications that have positive serology as a prerequisite for diagnosis are prone to both under- and overestimation of the incidence of rheumatoid arthritis. Rheumatoid factor seropositivity is more useful as a diagnostic than a screening criterion. Nonetheless, the presence of rheumatoid factor is not a necessary criterion for the diagnosis of this condition; nor is its absence adequate justification for the exclusion of rheumatoid arthritis in a patient with suggestive clinical findings. Rheumatoid factor is more often absent than present in the early stages of the disease and remains absent in one in ten diagnosed cases. On the other hand, patients with cirrhosis, sarcoidosis, and tuberculosis may be seropositive in the absence of clinical evidence of rheumatoid arthritis. The diagnosis of early rheumatoid arthritis is determined by clinical rather than laboratory assessment (8).

While the presence or absence of rheumatoid factor is neither a necessary nor a sufficient criterion on which to base a diagnosis of rheumatoid arthritis, it is a useful clinical prognostic indicator. A high-titer rheumatoid factor predicts a worse clinical course in a recently diagnosed case of rheumatoid arthritis. Persons with a high titer early in the course of their disease are at increased risk of developing rheumatoid nodules and erosive disease. Another laboratory test that is useful for monitoring disease progression is the erythrocyte sedimentation rate (ESR) (9). The ESR is regarded by the American College of Rheumatology as the most important laboratory criterion for determining remission or improvement in rheumatoid arthritis. Although an ESR cannot distinguish between rheumatic and other sources of inflammation, it is a reliable index of inflammatory activity. Sequential measures of the ESR provide a useful mode for objectively evaluating and recording exacerbations and remissions in rheumatoid arthritis patients. C-reactive protein, another acute-phase reactant, is used by some rheumatologists in preference to the ESR. In rheumatoid arthritis, laboratory investigations provide a more useful tool for monitoring progress and for excluding other causes of inflammatory polyarthritis than for diagnosing this disease (10). The same may be said of hyperuricemia in the diagnosis of gout.

Although urate deposition is generally a function of hyperuricemia and the risk of gout increases with a raised serum uric acid, hyperuricemia is not pathognomonic of gout. In fact, serum uric acid is considered a poor diagnostic tool as it frequently falls during acute attacks and may be normal; many people with hyperuricemia do not have gout (11). Gout results when urate crystals, deposited in joint tissues, provoke an acute inflammatory arthritis. The single most useful diagnostic criterion for gouty arthritis is the identification of urate crystals in joint fluid. Fluid aspirated from the inflamed joint of a patient with gout characteristically is cloudy with a gold or yellow color, contains needle-like crystals of monosodium urate, and has a high neutrophil count.

The first attack of gout is typically monoarticular involving the first metatarsophalangeal joint. The other anatomical areas most susceptible to the initial attack of gouty arthritis are, in order of reducing frequency, the instep, ankle, heel, wrist, finger, and elbow. The patient with gout presents with agonizing pain, which usually subsides, even without therapy, within three days. The patient remains relatively symptom-free, usually for six to 48 months, between the first and subsequent attack. The natural history of the disease involves attacks of increasing frequency and longer duration, with multiple joint involvement. After some years patients develop tophi, deposits of monosodium urate crystals in cartilage, synovial membranes, and soft tissue. Tophi occur as nodules in the helix of the ear, around the olecranon bursa, on the ulnar surface of the forearm, and on the fingers, hands, feet, knees, and Achilles tendon. The Arthritis Foundation uses the following criteria in the diagnosis of acute gouty arthritis (12):

- The presence of urate crystals in the aspirate from the acutely inflamed joint. Uric acid crystals demonstrate strong negative birefringence and are acicular or needle-shaped.
- Positive identification of urate crystals within a tophus.
- The presence of at least 50% of the following clinical, laboratory, or radiological criteria:
  - Two or more attacks of acute arthritis.
  - Inflammation that peaks within 24 hours.
  - A monoarticular arthritis.
  - A red joint.
  - A swollen or painful first metatarsophalangeal joint.
  - A unilateral attack involving the first metatarsophalangeal joint.
  - A unilateral attack involving a tarsal joint.
  - A suspected tophus.
  - Hyperuricemia.
  - Asymmetrical joint swelling evident on x-ray.
  - Radiological evidence of subcortical cysts in the absence of erosions.
  - A sterile joint fluid aspirate; the joint fluid should be aspirated and cultured during the acute attack.

A diagnosis of gout is based largely on designated criteria present in the patient; further qualification of the diagnosis as primary or secondary gout may rely on the absence of criteria in a given case. When the etiology is unknown, the absence of known criteria may determine the ultimate diagnostic label. A diagnosis of primary gout is made in the absence of myeloproliferative disorders, polycythemia, alcoholism, renal disease, drug therapy involving thiazide diuretics, cytotoxic agents, or low-dose aspirin, and various other conditions associated with hyperuricemia.

In overview, disease prevalence provides the general milieu in which clinical decision making is attempted. For the individual patient, the anatomical charting of joint involvement is considered a useful early differentiating feature in seeking to clarify the type of arthritis. Peripheral involvement is encountered in osteoarthritis, rheumatoid arthritis, psoriatic arthritis, septic arthritis, gout, and pseudogout; a central distribution is found in osteoarthritis, ankylosing spondylitis, septic arthritis, and degenerative disc disease. The involvement of metatarsophalangeal joints favors gout as a working hypothesis; interphalangeal joint involvement is more likely the result of rheumatoid arthritis, psoriactic arthritis, or Reiter's syndrome. In addition to the anatomical distribution of joint involvement, arthritis can be dichotomously classified according to the course of the disease. Acute arthritides include, amongst others, gout, septic arthritis, rheumatic fever, rheumatoid arthritis, psoriatic arthritis, Reiter's syndrome, and pseudogout. Chronic arthritis includes osteoarthritis, ankylosing spondylitis, tumor, and the last four conditions listed under the acute category. Natural histories, like anatomical categorizations, demonstrate marked overlap. Syndrome or pattern diagnosis attempts to overcome these overlaps by increasing the number of diagnostic criteria required before definitive labeling is attempted. This is the approach used to distinguish rheumatoid arthritis from psoriatic arthritis. A confident diagnosis of psoriatic arthritis can be made when: cutaneous psoriasis (silvery scales on a bright red plaque) precedes the onset of arthritis; articular and cutaneous lesions recur together and relapse simultaneously; distal rather than proximal interphalangeal joints are involved; and psoriatic nail stippling is detected. In a number of patients, however, the arthritis of psoriasis is largely indistinguishable from that of rheumatoid arthritis. In chronic cases, ulnar drift and subluxation would suggest rheumatoid arthritis; the presence of rheumatoid nodules, rheumatoid factor seropositivity, and periarticular osteopenia rather than fluffy periosteal new bone formation would also be helpful. No single strategy suffices in diagnostic labeling; an integrated approach is necessary.

Integration of diagnostic decision making strategies is well demonstrated in differentiating between gout and pseudogout. Probabilistic reasoning based on incidence and prevalence statistics suggests that for every two or three patients with gout, there is one patient with pseudogout. Acute gout, however, may coexist in almost one in ten cases of pseudogout. The distribution of anatomical lesions suggests that while gout, at some stage of its natural history, targets the metatarsophalangeal joint, pseudogout is more likely to involve the knee, wrist, shoulder, or ankle. All patients with gouty arthritis do suffer an acute attack; some patients with pseudogout have never had this experience. A definitive separation of gout and pseudogout may be achieved by aspirating involved joints and finding urate crystals in cases of gout and calcium pyrophosphate dihydrate crystal in cases of pseudogout. Aspiration of these crystals from involved joints is also a definitive criterion used in distinguishing between pseudogout and osteoarthritis, in which clinical and radiological findings are similar but the distribution of involved joints differs; and rheumatoid arthritis: one in ten patients with pseudogout have polyarticular involvement, and one in five have morning stiffness, synovial hyperplasia, and flexion contractures of the wrist, elbows, and knees.

Definitive diagnosis may rest on a single objective quantifiable criterion, or on the gestalt of the clinical picture. Syndrome diagnosis using a criterion-ranked format has also been developed for the confident pattern identification of pseudogout (13). Syndrome diagnosis is somewhat less definitive than identification of calcium pyrophosphate dihydrate crystals using either x-ray diffraction or chemical analysis following joint aspiration or biopsy. Pseudogout is probable when monoclinic or triclinic crystals showing no or weak positive birefringence on compensated polarized light microscopy are identified, or when typical calcifications are evident on x-rays of affected areas. Pseudogout is possible and should be suspected in patients with acute arthritis, particularly of the knees or other large joints, with or without hyperuricemia, or in those with chronic arthritis, especially of the knees, hips, wrists, carpi, elbows, shoulders, and metacarpophalangeal joints. Acute exacerbations further increase the likelihood of pseudogout.

Some may consider attaching an appropriate label to be an obsessive pastime in clinical health care. Certainly, every effort is made to precisely diagnose the presenting condition. The actual usefulness of diagnostic labeling, however, is a function of the nosological knowledge of the diagnosed condition's natural history and its documented response to intervention. The clinical value of the diagnostic labeling of a patient is directly proportional to the prognostic and therapeutic clues provided by a precise diagnosis. Gout, osteoarthritis, and rheumatoid arthritis all have detailed criterion-based syndrome descriptions, yet labeling in each of these conditions neither ensures an equivalent understanding of the underlying etiopathogenesis nor provides comparable guidance for specific intervention. In fact, it has been suggested that: "Gout is largely solved, both from diagnostic and therapeutic standpoints" (14); and, "Although OS [osteoarthritis] is recognized as the most common joint disorder, its pathogenesis is unclear and little is known about its natural history and progression" (15); and, "The cause of rheumatoid arthritis is unknown and therapy, therefore, remains empirical" (16). Syndrome diagnosis, while providing a detailed description of the disease state, does not inherently provide insight into the nature of the underlying disease process. Pattern diagnosis does not inevitably provide clear direction for specific syndrome management. Therapy for disorders diagnosed on the basis of their clinical patterns is not necessarily directed at modifying the etiology and pathogenesis of the condition; it may be directed solely at alleviating distressing symptoms. The extent to which therapy based on syndrome diagnosis is symptomatic depends on the nosological knowledge of the diagnosed condition. The management of a condition is no more rational than the understanding of its pathogenesis.

Knowledge of etiology makes disease prevention possible; an understanding of pathogenesis facilitates prognosis modification and the alteration of natural history. Acute gout with its sudden, severe, and relatively short (three to ten days) self-limiting attacks is the result of sodium monourate crystal deposition in joints. Gout may theoretically be prevented either by eliminating any identified causes or by modifying variables that permit formation or accumulation of urate crystals. A patient whose gout is secondary to hyperuricemia following persistent thiazide administration may be "cured" by ceasing ingestion of thiazide diuretics. In primary gout where the factor(s) initiating the condition are unclear, knowledge of pathogenesis makes specific intervention possible. Intervention in primary gout seeks to reduce urate crystal formation by reducing circulating levels of uric acid. Gout only occurs in instances of urate supersaturation; hyperuricemia, the metabolic precursor of gout, may be mild and asymptomatic.

An understanding of the mechanisms involved in the evolution of hyperuricemia in individual patients is an important stage in management decision making (17). Hyperuricemia may be reduced by decreasing uric acid production or increasing uric acid excretion. Most urate production is endogenous and related to the normal processes of cellular turnover; nonetheless, limitation of dietary purines (the chemical precursors of uric acid) can provide a useful management adjunct. Foods that should be avoided include meat, meat extracts, and offal; poultry and fish, especially sardines, herring, and anchovies; and yeast, alcohol, beans, peas, lentils, and spinach. In addition, uric acid formation may be blocked by drugs. Allopurinol blocks xanthine oxidase, the enzyme required for the conversion of hypoxanthine and xanthine to uric acid. This drug prevents urate supersaturation by blocking formation of uric acid. It does, however, lead to accumulation of xanthine. Xanthine is more soluble than uric acid and therefore less likely to precipitate in joints and other tissues. Side effects associated with the use of allopurinol include: exfoliative dermatitis, bone marrow depression, and hepatotoxicity. Another strategy for addressing hyperuricemia is to examine the excretion of uric acid. About two-thirds of urate is excreted by the kidney; the remaining third is excreted into the gut, where it is degraded by colonic bacteria. Modifications to urate excretion focus on the renal pathway. A daily urinary volume of greater than 1.5 liters is required to carry the normal (insoluble) urate load; an adequate fluid intake is essential in persons with hyperuricemia. Renal lithiasis is a significant problem. Urinary excretion of uric acid depends on the glomerular filtrate and renal tubular function. Reduced tubular excretion of uric acid is considered an important factor in hyperuricemia. In patients with normal renal function, renal excretion of urate can be enhanced using uricosuric agents such as salicylates and probenecid. Strategies that reduce the level of uric acid seek to prevent the disease. Intervention at this level requires a fairly complete understanding of the etiopathogenesis of a condition.

Intervention may also be based on a less comprehensive understanding of the mechanisms underlying the genesis of acute gout. Therapy may seek merely to modify the inflammatory process; colchicine and antiinflammatory agents are used with success in patients with acute gouty arthritis (18,19). Although such intervention is largely symptomatic and addresses patient distress rather than the cause of the disease, it does offer a realistic alternative when the knowledge of disease pathogenesis is incomplete. This strategy is, in fact, the basis of drug therapy in rheumatoid arthritis.

Rheumatoid arthritis is amenable to neither prevention nor cure. This situation is likely to prevail until the etiology of rheumatoid arthritis is clarified. Primary disease prevention becomes feasible when the etiology of a disease is both identifiable and avoidable. It is suspected that persons with a genetic predisposition to rheumatoid arthritis may respond to particular infections or other immune system triggers by initiating the inflammatory changes that regularly emerge during the pathogenesis of the disease (20). Management in rheumatoid arthritis may therefore focus either on modifying the patients's inflammatory response or eliminating rheumatoid arthritis triggers. The identity of such triggers is obscure, although certain authors have suggested that food intolerance may, in a certain group of patients, constitute an immune trigger in rheumatoid arthritis (21). In the case of management through modification of the inflammatory response, and understanding of inflammatory mechanisms provides a number of possible nonspecific interventions. In attempts at management by eliminating triggers, the uncertain nature of these etiological triggers provides a daunting challenge to disease prevention.

The central role played by inflammation in the pathogenesis of rheumatoid arthritis is well recognized and the subject of intense investigation and speculation. Histological changes accompanying the clinical manifestations of rheumatoid arthritis include proliferation of synovial lining cells and joint infiltration by lymphocytes, plasma cells, macrophages, and polymorphs. In time, an ingrowth of granulation tissue over the perichondral margins spreads to form a pannus over the articular surfaces, with consequent progressive joint deformity. Active synovitis appears to be the main cause of cartilage loss (22). The decreased production of suppressive B cell factor by the B lymphocytes of synovial membranes may contribute to the perpetuation of rheumatoid factor and chronic synovial inflammation (23). A second postulate suggests that locally produced antibodies complex with unidentified inciting antigens, thereby activating complement (24); the subsequent inflammatory cascade involving primarily polymorphs and their products may induce both the inflammatory and the destructive changes characteristic of rheumatoid arthritis. "Complement activation is recognized as a central event in inflammation in the rheumatoid joint" (25). Monocytes are also hypothesized to be involved in the pathogenesis of this complex condition. Peripheral monocytes isolated from patients with rheumatoid arthritis have an enhanced capacity to generate proinflammatory metabolites of arachidonic acid (26). Arachidonic acid metabolites are among the biochemical mediators contributing to the morphological changes in an arthritic joint.

The products of arachidonic acid metabolism enhance both humeral and cellular responses in inflammation and modify the local immune response.

Arachidonic acid when exposed to cyclooxygenase generates series 2 prostaglandins; exposure of arachidonic acid to lip oxygenase results in the generation of series 4 leukotrienes (LTB$_4$). Both groups of products, to varying degrees, enhance inflammation. Leukotrienes promote leukocyte chemotaxis, enhance release of lysosomal enzymes, increase vascular permeability, and are markedly proinflammatory. Series 2 prostaglandins include thromboxanes (TXA$_2$), prostacyclin, and prostaglandin E$_2$ (PGE$_2$). Depending on the tissue involved, the major product of arachidonic acid metabolism varies (27); in the joints of rheumatoid arthritis patients, PGE$_2$ is of great importance. This compound has been shown, at low concentrations, to cause erythema and potentiate the edema and pain resulting from inflammatory mediators such as bradykinin and histamine (28). At physiological concentrations, PGE$_2$ has been shown in vitro to suppress T lymphocyte functions and inhibit the cytotoxicity of non-T lymphocytes. In fact, it has been hypothesized that endogenous production of PGE$_2$, through tonic inhibition of suppressor cell activity, by enhancing autoantibody production results in the accumulation of rheumatoid factor in affected joints (29). Owing to the short half-life of PGE$_2$, this compound's activity is largely limited to its site of production. Regardless of whether the major effect of PGE$_2$ is through its influence on the immune system or its direct, or more importantly indirect, proinflammatory effect, a high concentration of cyclooxygenase-derived products is associated with enhanced inflammation in a joint; reduction of these products is followed by reduced inflammation (30).

A tentative model linking rheumatoid arthritis and arachidonic acid metabolism has been proposed. There is some clinical support for such a notion. The pathological explanation of this model proposes that the macrophages and leukocytes infiltrating the synovium increase the joint's concentration of PGE$_2$, TXA$_2$, and LTB$_4$. These products, assisted by activated fragments of complement, further stimulate the influx of leukocytes, which in turn aggravate inflammation by enhancing chemotaxis and local vascular permeability. PGE$_2$ is considered a particular villain as it is believed to mediate bone resorption by increasing osteoclast numbers, to stimulate collagenase secretion by macrophages, and to inhibit proteoglycan production by synoviocytes and articular chondrocytes. In vitro experimentation with cultures of rheumatoid synovial tissue supports the view that prostaglandins play a substantial role in the development of cartilaginous lesions in inflammatory arthritis (31, 32). Injured tissue generates inflammatory prostaglandins, chemotactically attracted leukocytes release prostaglandins during phagocytosis, and macrophages, the dominant cells in chronic inflammation, also produce prostaglandins. This inflammatory cycle could be self-perpetuating. Although aspects of the inflammatory process in the genesis of rheumatoid arthritis remain unclear (33), a logical management strategy for this disease may seek to modify arachidonic acid metabolism. This may be achieved by drugs or by dietary means.

Steroids and nonsteroidal antiinflammatory drugs (NSAIDs) are routinely considered in the pharmacological management of rheumatoid arthritis. The affect of both of these drug groups is mediated, at least in part, by their impact on arachidonic acid metabolism. Corticosteroids impair the release of arachidonic acid from cell membranes, thereby effectively inhibiting the synthesis of inflammatory products from arachidonic acid (34). Clinically, steroids are regarded as more effective than NSAIDs—they are used with greater caution, however, owing to their increased side effects. Salicylates and other NSAIDs, initially alone and later in combination, are the first- and second-line drug therapies recommended by certain rheumatologists (35). The manner in which these drugs exert their antiinflammatory effect is somewhat disputed depending on the researcher's training, but does seem to involve inhibition of the synthesis of prostaglandins (36). Cyclooxygenase, the enzyme required for the conversion of arachidonic acid to prostaglandins, is inhibited by NSAIDs; it has been suggested that the immediate symptomatic relief of chronic inflammation is the result of blocking cyclooxygenase-mediated metabolism.

Dietary intervention in the management of rheumatoid arthritis through modification of arachidonic acid metabolism, requires the patient's conversion and lifelong adherence to a diet rich in linoleic and linolenic acids rather than arachidonic acid. Arachidonic acid, usually esterified to membrane phospholipids and less frequently to triglycerides and cholesterol, is an important component of cell membranes. Cellular arachidonic acid is derived from the diet. Rich sources of preformed arachidonic acid include meat and animal products; dietary linoleic acid, an essential omega-6 fatty acid, is also an arachidonic acid precursor. Both of these omega-6 fatty acids, along with linolenic acid, an essential omega-3 fatty acid, can act as precursors of prostaglandins. Prostaglandins, depending on the series, may enhance or retard inflammation; their role as intracellular regulators of metabolism includes temperature modification, platelet aggregation, vascular contraction, and modulation of the immune response.

The concentration of the various types of synthesized prostaglandins and leukotrienes is influenced by the quantity and type of dietary unsaturated fatty acids. Important dietary sources of linoleic acid are vegetable oils; safflower seed oil contains over 70% linoleic acid, and soybean, corn, and cottonseed oils contain about 50%. Barley, wheat, and apples are also good sources of linoleic acid. In vivo conversion of linoleic acid to arachidonic acid depends on the presence of two desaturase enzymes. Δ-6-desaturase converts linoleic acid to γ-linoleic acid, the precursor of dihomo-γ-linoleic acid (Fig. 3.1). The latter may be converted into arachidonic acid or it may generate series 1 prostaglandins. Substrate competition for this rate-limiting enzyme in arachidonic acid synthesis may also occur from omega-3 fatty acids. α-Linolenic acid, the essential fatty acid of the omega-3 series, is the precursor of eicosapentaenoic acid (EPA), the substrate for the synthesis of the series 3 prostaglandins (Fig. 3.1). Series 1 and series 3 prostaglandins lack the proinflammatory activity of series 2 prostaglandins. Alteration of the substrate or the activity of these desaturase enzymes may therefore, in theory, modify the local inflammatory response.

Dietary manipulation may achieve alterations in these enzyme activities. The activity of Δ-6-desaturase, thought to be the rate-limiting enzyme in these reactions, is reduced by fasting, glucose, fructose, glycerol, and linoleic and arachidonic acids; its activity is enhanced by a protein-rich or essential

48 / DIAGNOSTIC DECISION MAKING IN CLINICAL PRACTICE

```
         OMEGA-6 SERIES                          OMEGA-3 SERIES

        CIS-LINOLEIC ACID                        α-LINOLENIC ACID
         (vegetable oils)                    (leafy vegetables, linseed oil)
                │                                       │
                │         Δ-6-Desaturase                │
                │  (rate limiting; favors omega-3 series)│
                ▼                                       ▼
       DI HOMO-γ-LINOLEIC ACID               EICOSAPENTAENOIC ACID
              │  ╲                                  (fish oils)
              │   ╲                                    │    ╲
              │    ▼                                   │     ╲
              │   SERIES 1                             │      ▼
              │  PROSTAGLANDINS                        │    SERIES 3
              │  anti-inflammatory                     │  PROSTAGLANDINS
              │                                        │  weakly inflammatory
              ▼                                        ▼
        ARACHIDONIC ACID                      DOCOSAHEXAENOIC ACID
          (meat, eggs)                              (fish oils)
              │                                    ╱
              │                                   ╱
              │          Membrane Store          ╱
              │                                 ╱
  Phospholipase A2       ┤────── Steroids
              │                  Bioflavonoids
              ▼
      FREE ARACHIDONIC ACID
         ╱          │           ╲
  Lipoxygenase    NSAIDS ──┤    Cyclooxygenase
       ╱                          ╲
      ▼                            ▼
  LEUKOTRIENES               PROSTAGLANDIN E2
  strongly inflammatory         inflammatory
```

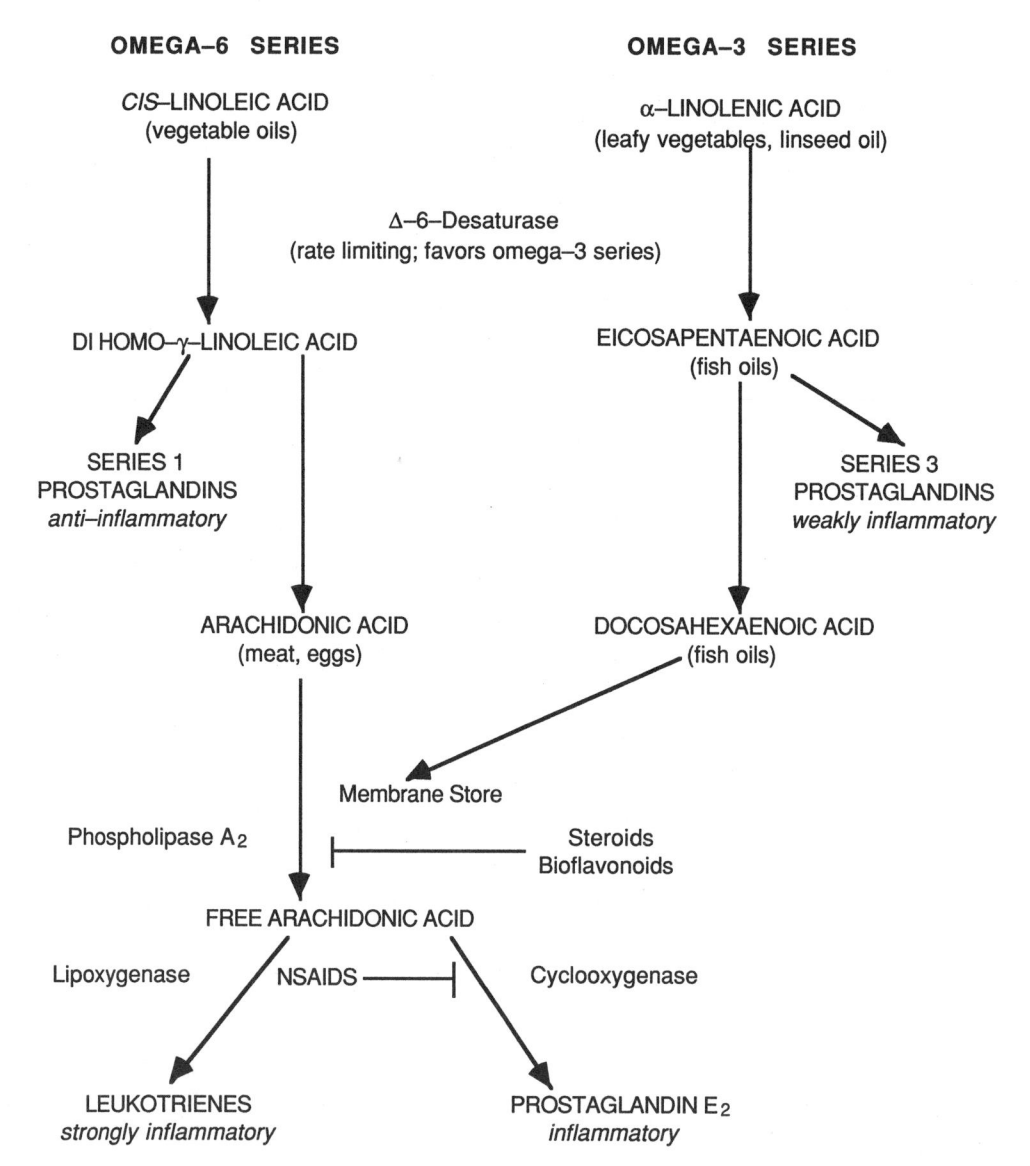

Figure 3.1. Arachidonic acid and the generation of inflammatory mediators. *NSAIDS*, nonsteroidal antiinflammatory drugs.

fatty acid–deficient diet (37). Linolenic acid competes successfully with linoleic acid for Δ-6-desaturase. Dietary sources of α-linolenic acid are linseed oil, spinach, beans, and aquatic plants.

In addition to attempting to alter the production of prostaglandins by targeting desaturase enzyme activity, it is also possible to enhance the production of the series 3 prostaglandins by supplementing the diet with foods rich in omega-3 fatty acids. Cod, mackerel, herring, crab, shrimps, and oysters along with other cold-water fish are sources of EPA. Such a dietary change from omega-6 to omega-3 fatty acid sources is postulated to reduce the production of the proinflammatory and platelet-aggregatory series 2 prostaglandins and the leukotriene products of lipoxygenase activity. Incorporation of EPA into cellular membranes by dietary supplementation is postulated to: inhibit the cyclooxygenase pathway leading to production of the series 2 prostaglandins; produce leukotrienes that are ten to 30 times less potent than those resulting from arachidonic acid catabolism by lipoxygenase; and alter the function of human leukocytes (38). This concept has been applied clinically.

Dietary changes can alter the balance of various essential fatty acids in cell membranes. Clinical investigation has shown that dietary supplementation of up to 10 g/day with eicosapentaenoic acid can increase the EPA content of red cell and platelet phospholipids (39). Lee and coworkers (40) demonstrated that by supplementing the diet with marine oils (EPA and docosahexaenoic acid) they could alter the lipid composition of neutrophil and monocyte membranes. They showed that such an alteration had an overall antiinflammatory effect in vitro, and postulated that this was attributable to the suppression of the lipoxygenase pathway in both types of cells, complemented by the inhibition of LTB_4-mediated functions in neutrophils. A subjective alleviation of active rheumatoid arthritis by altering essential fatty acid intake has been achieved clinically in two recent double-blind studies. A non-crossover study used fish oil supplementation in doses of 18 g/day (41); a controlled crossover study used daily MAXEPA administration of 2.7 g of EPA plus 1.8 g of docosahexaenoic acid (42). Animal studies suggest that the symptomatic use of fish oils in rheumatoid arthritis may be extrapolated to other arthritides associated with collagen disorders (43). It should be emphasized that arachidonic acid metabolites are only one set of a number of inflammatory mediators (44). This area requires more research, which should take into account both the effectiveness of such intervention and any side effects that may result from prolonged therapy.

Dietary intervention as a strategy in the management of rheumatoid arthritis is not without its critics. One valid criticism points out the difficulties encountered in appropriately interpreting clinical change in a disease that has a natural history of exacerbations and remissions (45). Even interpretation of the pathogenic significance of a detected clinical change may be difficult. A diet high in polyunsaturated fats, which limited saturated fats and used an EPA supplement, MAXEPA, was offered to 17 patients with rheumatoid arthritis (46). After 12 weeks on this diet the experimental group suffered less morning stiffness and joint tenderness than the controls. Two months after stopping the diet the experimental group showed clinical deterioration. It has been suggested that, rather than indicating a response to altered fatty acid metabolism, this study demonstrates the unmasking of a food allergy following 12 weeks on an elimination diet (47). Some anecdotal support for this type of explanation may be sought in the experimental diet used by Panush and associates (48). These authors used a "popular" intervention diet—i.e., one free of preservatives, additives, red meats, dairy products, herbs, and fruit—in the management of their rheumatoid arthritis patients. The results failed to demonstrate objective overall improvement in the experimental group; nonetheless, marked improvement in two patients did lead the authors to propose that selected rheumatoid arthritis patients may benefit from dietary manipulation. The possibility that the patients who benefited from the experimental diet were food intolerant and that the documented improvement resulted from the "elimination diet" deserves further investigation.

Type I (reaginic), type III (immune complex), and type IV (cell-mediated) hypersensitivity have all been linked to food allergy (49). At least one of these reactions has been regularly implicated in rheumatoid arthritis. Immune complexes are clearly a basic element in the pathogenesis of this disease (50). The classic clinical triad of food allergy is currently confined to dermatological, gastrointestinal, and respiratory manifestations—it does not at present extend to incorporate joint lesions. A more contemporary view of food intolerance that does not adhere to the traditional criteria of food allergy does recognize arthralgia and arthritis as potential outcomes of chronic exposure to an allergen in an individual with cyclical food allergy (51). Anecdotal evidence and a limited number of case studies certainly suggest that this possibility should be seriously considered (52).

If food allergy is ultimately deemed to constitute a valid trigger to rheumatoid arthritis, this information has implications for the antiinflammatory drug therapy currently prescribed in rheumatoid arthritis. Arthritic patients taking NSAIDs have been shown to have increased intestinal permeability (53). If food allergy is a trigger to rheumatoid arthritis, drugs currently considered routine medication may perpetuate the condition in susceptible patients. Intervention now considered to have an acceptable cost–benefit ratio, given the current understanding of the etiopathogenesis of rheumatoid arthritis, may be contraindicated at a future date.

Advances in the management of rheumatoid arthritis are expected to parallel a better understanding of the etiopathogenesis of this condition (54). Currently, dietary therapy by means of altering arachidonic acid metabolism, like many other nontraditional treatments in the management of rheumatoid arthritis, remains unproven (55). Identification of, and dietary management to avoid, food allergen exposure, although presently unproven, may one day constitute a valid clinical strategy in appropriate patients (56). Therapy that is now routinely prescribed for rheumatoid arthritis may one day be regarded as carrying an unacceptable risk. These are the dilemmas encountered when intervention is based on an inadequate understanding of the etiopathogenesis of a condition.

A similar problem confronts the management of osteoarthritis. Current treatment "of this very common arthropathy has yet to be specifically directed at [a] key pathogenic mechanism" (57). Reliance on nonspecific intervention in the management of osteoarthritis is testimony to the lack of definite information, despite a multitude of theories, on the genesis of this condition. Unfortunately, the management of osteoarthritis can be no more rational than the understanding of its natural history. "The etiology of osteoarthritis, the most common articular disorder, is still unknown" (58). In fact, there is growing consensus that osteoarthritis is a label used to describe a heterogeneous set of diseases that share a final common pathway (15). Surveys of various anatomical regions showing osteoarthritic changes suggest several clinicoradiological subsets, a phenomenon that may be the result of different pathophysiological mechanisms (59). This postulated explanation of secondary osteoarthritis, which may be encountered in rheumatoid arthritis, pseudogout, diabetes, and hemophilia, is also applicable to primary osteoarthritis.

Primary osteoarthritis is categorized as idiopathic and generalized; an inflammatory variant is termed erosive osteoarthritis (57). While the "idiopathic" nomenclature acknowledges the presence of unrecognized etiopathological mechanisms, the genesis of generalized osteoarthritis is itself fraught with contradictions. The major risk factor for generalized osteoarthritis is believed to be a genetic predisposition; cartilage vulnerability is thought to be genetically determined. Given generalized cartilage susceptibility, site specificity is thought to result from the interaction of diverse variables. Osteoarthritis of the knee correlates with injury, repetitive use, and obesity (60); osteoarthritis of the hip is associated with dysplasia and faulty biomechanical stress loading (61). The correlation of age with osteoarthritis may also be explained by the accumulated stress hypothesis. While mechanical overloading is recognized as an important variable in the genesis of osteoarthritis, it remains to be seen whether work or leisure sports are the major contributors to the prevalence of this condition (62). Anecdotal observations linking recreational sports to osteoarthritis have not necessarily been confirmed by the few well-controlled trials reported (63). Although suggestions that excessive exercise may damage cartilage cannot be discounted, it may be prudent to remember that disuse also results in cartilage deterioration. Clarification of the nature and extent of exercise congruent with healthy joints requires further investigation.

The final common pathway in the pathogenesis of osteoarthritis involves cartilage loss. Cartilage contains some cells—chondrocytes—and an abundant matrix, which is 90% water; other major matrix constituents are collagen and proteoglycans. Proteoglycan degradation is believed to be central to the development of osteoarthritis (64). Proteoglycans, which consist of glycosaminoglycans (particularly chondroitin and keratin sulfate) attached to a protein core, may be degraded by lysosomal enzymes released from chondroctyes, synoviocytes, or leukocytes. Degradation of proteoglycans results in loss of their normal negative charge and impairment of their hydrophilic character, with consequent deterioration in joint deformability and elasticity. There is a direct correlation between the severity of the disease and the reduction in proteoglycans. Osteoarthritis does not appear to be the result of diminished cartilage metabolism; on the contrary, matrix synthesis and cell replication proceed at greater rates in damaged than in normal cartilage (65). Furthermore, the response of articular chondrocytes does not appear to diminish with age (66). The "inadequacy of the chondrocyte response" hypothesis would thus appear less appropriate than the "excessive stress imposed upon normal tissue" postulate. The genetic susceptibility of certain individuals may be expressed not as a defect in the ability of chondrocytes to produce quantities of matrix, but rather as an inability to reverse cartilage fibrillation in the presence of excessive matrix catabolism. In instances of inflammatory osteoarthritis, the predominant stress may be biochemical; in the majority of cases the initiating stress may be largely mechanical. Biomechanical stress may certainly be increased in instances of developmental joint abnormalities, trauma, imbalances in joint protective mechanisms, and/or modification to tissue shock absorption (67). Osteoarthritis may largely be an expression of underlying biomechanical imbalance. Like cancer, it may only lead to overt disease when initiating factors, such as mechanical stress in the case of osteoarthritis, are expressed in the presence of a promoting factor, the genetic predisposition. Speculation abounds, but the etiopathogenesis of osteoarthritis remains elusive. Until such time as this etiopathogenesis is clarified, treatment will be nonspecific. Current therapy focuses on symptomatic problems; physical therapy, antiinflammatory drugs, and, in advanced cases, surgery are used. Only a knowledge of the etiopathogenesis of osteoarthritis will provide the options of disease prevention and definitive therapy.

Syndrome diagnosis provides a good clinical approach to diagnostic labeling. Pattern analysis is thought to generate more precise diagnoses and prognostication as information is handled as data sets rather than as discrete data collections (68). Pictorially represented by Venn diagrams, syndrome diagnosis is a useful means of categorizing patients; the prerequisite for such an approach is a good description of the clinical features of the analyzed condition.

Syndrome diagnosis, while requiring and ensuring a clear description of clinical manifestations, does not necessarily imply an acceptable understanding of disease genesis and/or intervention. It would be naive to assume that a well-defined set of criteria for recognizing a condition automatically implies a good understanding of how to prevent and/or treat that condition. Syndrome diagnosis may comprehensively clarify the nuances of clinical presentation in diagnostic decision making, but rational management decisions achieve better clinical outcomes when based on cause–effect relationships.

References

1. Demlow LL, Liang MH, Eaton MH: Impact of chronic arthritis in the elderly. *Clin Rheum Dis* 12:329–335, 1986.
2. Weiss TE, Biundo JJ: Rheumatic diseases. *Postgrad Med* 60:141–150, 1976.
3. Baum J: Joint pain. *Postgrad Med* 85:311–321, 1989.
4. Brand C, Barraclough D: Osteoarthritis. *Curr Ther* Aug:51–61, 1985.
5. Calin A: Degenerative joint disease. *Am Fam Physician* 33:167–172, 1986.
6. Bjarnason D, Dlesk A, et al: Early rheumatoid arthritis. *Postgrad Med* 79:47–61, 1986.
7. Ropes MW, Bennett GA, et al: 1958 revision of diagnostic criteria for rheumatoid arthritis. *Bull Rheum Dis* 9:175–176, 1985.
8. Milazzo S: Assessment and management of the difficult arthritic. *Patient Manage* Oct:75–82, 1979.
9. Boulware DW, Kushner I: Rheumatologic tests: what to expect. *Patient Care* July:164–186, 1989.
10. Brand C, Muirden KD: The place of investigations in rheumatoid arthritis. *Patient Manage* June:19–33, 1988.
11. Henderson D: Gout. *Aust Fam Physician* 15:1276–1282, 1986.
12. Wallace SL, Robinson H, Masi AT, et al: Preliminary criteria for the classification of acute arthritis of primary gout. *Arthritis Rheum* 20:895–900, 1977.
13. McCarty DJ: Calcium pyrophosphate dihydrate crystal deposition disease. In Schumacher

HR (ed): *Primer on the Rheumatic Diseases.* Atlanta, Arthritis Foundation, 1988, p 209.
14. McCarty DJ: Crystal identification in human synovial fluids: methods and interpretations. *Rheum Dis Clin North Am* 14:253–267, 1988.
15. Davis MA: Epidemiology of osteoarthritis. *Clin Geriatr Med* 241–255, 1988.
16. Dugowson CE, Gilliland BC: Management of rheumatoid arthritis. *Dis Mon* 32:9, 1986.
17. Becker MA: Clinical aspects of monosodium urate monohydrate crystal deposition disease (gout). *Rheum Dis Clin North Am* 14:377–394, 1988.
18. Buchanan WW, Kean W: Hyperuricemia and gouty arthritis. *Mod Med Aust* 30:64–73, 1987.
19. Fox IH, Koopman WJ, Schumacher HR: Gout and pseudogout in primary care. *Patient Care* July:84–126, 1989.
20. Hughes GRV. Rheumatoid arthritis. *Br J Hosp Med* June:584–592, 1979.
21. Darlington LG: Does food intolerance have any role in the aetiology and management of rheumatoid arthritis? *Ann Rheum Dis* 44:801–804, 1985.
22. Shiozawa S, Shiozawa K: A review of the histopathological evidence on the pathogenesis of cartilage destruction in rheumatoid arthritis. *Scand J Rheumatol Suppl* 74:65–72, 1988.
23. Pisko EJ, Skipper ER, Foster SL, Panetti M, Challa VR, Geisinger KR, et al: Decreased production of suppressive-B-cell factor by synovial membrane B-lymphocytes in rheumatoid arthritis. *Clin Exp Immunol* 73:473–478, 1988.
24. Zvaifler NJ: New perspectives on the pathogenesis of rheumatoid arthritis. *Am J Med* 85:12–17, 1988.
25. Morgan BP, Daniels RH, Williams BD: Measurement of terminal complement complexes in rheumatoid arthritis. *Clin Exp Rheumatol* 6:239–245, 1988.
26. Colli S, Caruso D, Stragliotto E, Morazzoni G, Aletti A, Galli G, et al: Proinflammatory lipoxygenase products from peripheral mononuclear cells in patients with rheumatoid arthritis. *J Lab Clin Med* 112:357–362, 1988.
27. Gurr MI: *Role of Fats in Food and Nutrition.* London, Elsevier Applied Science Publishers, 1986, pp 94–108.
28. Regtop H: How essential are essential fatty acids? *Int Clin Nutr Rev* 3:5–18, 1983.
29. Goodwin JS: Immunologic effects of nonsteroidal anti-inflammatory agents. *Med Clin North Am* 69:793–804, 1985.
30. Kuehl FA, Egan RW: Prostaglandins, arachidonic acid and inflammation. *Science* 210:978–984, 1980.
31. Lippiello L, Yamamoto K, Robinson D, Mankin HJ: Involvement of prostaglandins from rheumatoid synovium in inhibition of articular cartilage metabolism. *Arthritis Rheum* 21:909–916, 1978.
32. Robinson DR, Dayer JM, Krame SM: Prostaglandins and their regulation in rheumatoid inflammation. *Ann N Y Acad Sci* 332:279–294, 1979.
33. Kolarz G: Pathogenesis of rheumatoid arthritis. *Acta Med Austriaca* 15:128–130, 1988.
34. Metz SA: Anti-inflammatory agents as inhibitors of prostaglandin synthesis in man. *Med Clin North Am* 65:713–754, 1981.
35. Roth SH: Arthritis therapy: how to get through the maze of options. *Mod Med Aust* Jan:14–21, 1985.
36. Hart FD, Huskisson EC: Non-steroidal anti-inflammatory drugs—current status and rational therapeutic use (part I). *Curr Ther* May:97–113, 1984.
37. Brenner RR: Nutritional and hormonal factors influencing desaturation of essential fatty acids. *Prog Lipid Res* 20:41–47, 1982.
38. Krilis SA: Leukotrienes as mediators of disease. *Patient Manage* Sept:123–130, 1986.
39. Knapp HR, Reilly IAG, et al: In vivo indexes of platelets and vascular function during fish oil administration in patients with atherosclerosis. *N Engl J Med* 31:937–942, 1986.
40. Lee TH, Hoover RL, et al: Effect of dietary enrichment with eicosapentaenoic and docosahexaenoic acids on in vitro neutrophil and monocyte leukotriene generation and neutrophil function. *N Engl J Med* 312:1217–1224, 1985.
41. Cleland LG, French JK, Betts WH, Murphy GA, Elliot MJ: Clinical and biochemical effects of dietary fish oil supplements in rheumatoid arthritis. *J Rheumatol* 15:1471–1475, 1988.
42. Kremer JM, Jubiz W, Michalek A, Rynes RI, Bartholomew E, Bigaquette J, et al: Fish oil fatty acid supplementation in active rheumatoid arthritis. *Ann Intern Med* 106:497–503, 1987.

43. Leslie CA, Conte JM, Hayes KC, Carthcart ES: A fish oil diet reduces the severity of collagen induced arthritis after onset of the disease. *Clin Exp Immunol* 73:328–332, 1988.
44. Snyderman R: Mechanisms of inflammation and leukocyte chemotaxis in the rheumatic diseases. *Med Clin North Am* 70:217–236, 1986.
45. Podell RN: Nutritional treatment of rheumatoid arthritis. *Postgrad Med* 77:65–72, 1985.
46. Kremer JM, Bigauoette J, et al: Effects of manipulation of dietary fatty acids on clinical manifestations of rheumatoid arthritis. *Lancet* 1:184–187, 1985.
47. Birtwistle S, McEwen LM: Letter to the *Lancet*. *Lancet* 1:700, 1985.
48. Panush RS, Carter RL, et al: Diet therapy for rheumatoid arthritis. *Arthritis Rheum* 26:462–471, 1983.
49. Stern MS, Walker WA: Food allergy and intolerance. *Pediatr Clin North Am* 32:471–492, 1985.
50. Williams RC: Immune complex-mediated rheumatic diseases. *Postgrad Med* 68:124–131, 1980.
51. Randolph TG: Ecologically orientated rheumatoid arthritis. In Dickey LD (ed): *Clinical Ecology*. Springfield, IL, Charles C Thomas, 1976, pp 249–255.
52. Panush RS, Webster EM: Food allergies and other adverse reactions to foods. *Med Clin North Am* 69:533–546, 1985.
53. Bjarnason I, So A, et al: Intestinal permeability and inflammation in rheumatoid arthritis: effects of non-steroidal anti-inflammatory drugs. *Lancet* 2:11, 1984.
54. Konttinen YT, Honkanen VE: Future trends in the treatment of rheumatoid arthritis in the light of current etiopathogenetic theories. *Scand J Rheumatol Suppl* 74:7–17, 1988.
55. Hawley DJ: Non-traditional treatment of arthritis. *Nursing Clin North Am* 19:663–672, 1984.
56. McCrae F, Veerapen K, Dieppe P: Diet and arthritis. *Practitioner* 230:359–361, 1986.
57. Shuckett R, Lee P: The pathogenesis of osteoarthritis. *Mod Med Aust* May:16–22, 1986.
58. Moskowitz RW: Primary osteoarthritis: epidemiology, clinical aspects and general management. *Am J Med* 83(5A):5–10, 1987.
59. Peyron JG: Osteoarthritis. The epidemiological viewpoint. *Clin Orthop* 213:13–19, 1986.
60. Felson DT: Epidemiology of hip and knee osteoarthritis. *Epidemiol Rev* 10:1–28, 1988.
61. O'Brein TM, Moran R, McGoldrick F: The aetiology of degenerative disease of the hip. A review of 400 cases. *Ir J Med Sci* 158:63–66, 1989.
62. Genti G: Occupation and osteoarthritis. *Baillieres Clin Rheumatol* 3:193–204, 1989.
63. Panush RS, Brown DG: Exercise and arthritis. *Sports Med* 4:54–64, 1987.
64. Clyne MJ: Pathogenesis of degenerative joint disease. *Equine Vet J* 19:15–18, 1987.
65. Bora FW, Miller G: Joint physiology, cartilage metabolism, and the etiology of osteoarthritis. *Hand Clin* 3:325–336, 1987.
66. Hough AJ, Webber RJ: Aging phenomena and osteoarthritis: cause or coincidence? Claude P. Brown memorial lecture. *Ann Clin Lab Sci* 16:502–510, 1986.
67. Radin EL, Paul IL, Rose RM: Osteoarthritis as a final common pathway. In Nuki G (ed): *The Aetiopathogenesis of Osteoarthritis*. London, Pitman Medical, 1980, pp 84–88.
68. Knoebel SB: Perspectives on clinical decision-making: the application of pattern analysis. *PACE* 11:2077–2081, 1988.

4

Causal Diagnosis: Unproven Pathogenesis, Uncertain Intervention

The value of a causal diagnosis is most apparent when the pathogenesis of a condition is proven; determination of the diagnostic label is tantamount to therapy selection. A patient who complains of lassitude and appears pale may be suffering from a diversity of conditions; a low hemoglobin and/or erythrocyte count would suggest anemia as a reasonable working diagnosis. Using a causal strategy, clinical evidence to substantiate or refute an anemia attributable to blood loss, hemolysis, bone marrow depression, or defective hemopoiesis is sought. Patients with anemia attributable to blood loss may have a history of menorrhagia; those with hemolytic anemia may have a tinge of jaundice; those with bone marrow depression may complain of easy bruising and susceptibility to infections. Laboratory investigations are particularly helpful in pinpointing the type of anemia present. While all anemias have a reduced hemoglobin level, only iron deficiency anemia has a microcytic hypochromic red cell morphology associated with depleted body iron stores, a low serum ferritin, and an increased serum total iron binding capacity. Patients with low hemoglobin levels, megaloblastic red cell morphology, and low red cell folate levels may be folate deficient; they may also lack adequate vitamin B_{12}. Further laboratory investigations that ascertain serum levels of these vitamins are obligatory. Accurate causal diagnosis is necessary as a prelude to appropriate therapy. Folic acid supplementation of a vitamin B_{12}-deficient individual will result in an improved blood picture and mask progressive combined degeneration of the spinal cord. Syndrome pattern recognition may be utilized in an attempt to refine the working hypothesis. Once a causal diagnosis has been established, the therapy attempts to eliminate, correct, or modify the cause. Causal diagnosis not only implies identification and correction of the precipitating cause, it also requires modification of other initiating and contributing factors. The diagnosis of persons as having iron deficiency anemia represents interim diagnostic labeling; further investigation to identify the reason for this anemic state is an integral component of comprehensive causal diagnosis. Only further data collection can discriminate between iron deficiency anemia attributable to dietary aberration and that attributable to overt or occult blood loss. The causal diagnosis of an individual with iron deficiency anemia may be bowel cancer or adherence to a restrictive vegetarian diet. Causal diagnosis, in optimal circumstances, implies identification of all the variables contributing to a disease state; it requires a comprehensive understanding of the genesis of a condition and provides clear therapeutic guidance.

Osteopenia, like anemia, is a nonspecific term that may serve as an interim anatomical and/or sign-based diagnosis. Further diagnostic decision making requires that a causal diagnosis be attempted. Osteopenia, a well-recognized aging phenomenon, may present clinically as osteoporosis, osteomalacia, osteitis fibrosa cystica, or other hormonal imbalances, or as a malignancy.

1. Osteoporosis. Osteoporosis is a clinically defined condition that is diagnosed when osteopenic bones fracture in the absence of significant trauma. Such patients have a reduced amount of bone per unit volume; a reduced amount of normally mineralized osteoid is present. The two major varieties of osteoporosis are categorized as senile and postmenopausal osteoporosis. Different types of bone loss have been associated with each of these categories. Senile bone loss is a gradually progressive phenomenon that starts insidiously when peak bone mass has been attained; typically, the predominantly cortical bone loss is detected clinically after 75 years of age (1). In postmenopausal bone loss, which is particularly associated with loss of trabecular bone density, accelerated bone loss occurs following menopause. After eight or ten years during which 2–3% of bone mass is lost annually, bone loss reverts to the original slow rate of the premenopausal period. Hormonal deprivation seems to particularly affect trabecular bone; some authors believe that trabecular spinal bone loss may reach 8% per annum following oophorectomy. The male:female ratio of hip fractures is 1:2, while that of vertebral fractures is 1:8 (2).

2. Osteomalacia. This condition is associated with an absolute or relative deficiency of vitamin D. Its contribution to osteopenia in the elderly is associated with impaired calcium absorption, largely attributable to faulty conversion of the partially active 25-hydroxy form of vitamin D to the fully active 1,25 dihydroxy form. Lack of calcium and/or phosphate for mineralization of bone matrix results in a decrease in the unit density of bone without any alteration in the size of either trabecular or cortical bone. The diagnosis can be confirmed by bone biopsy following tetracycline labeling.

3. Osteitis fibrosa cystica. This is the bone disease of primary hyperparathyroidism. Patients complain of thirst, polyuria, and constipation. Renal calculi may present as renal colic. Laboratory tests show a high serum calcium, low serum phosphate, elevated serum immunoreactive parathyroid hormone, and elevated urinary cyclic AMP. Secondary and tertiary hyperparathyroidism have a similar impact on bone metabolism.

4. Other hormonal imbalances. Osteopenia may also be caused by glucocorticoid and/or thyroxin excess. Cushing's syndrome is often iatrogeni-

cally induced by prolonged steroid therapy. The patient has wasted extremities, a round red face, and large abdomen/trunk. In patients complaining of weight loss despite a hearty appetite, and having fine skin and hair, heat intolerance, diarrhea, and muscle weakness, thyrotoxicosis should be suspected.

5. Malignancy. Malignancies such as multiple myeloma should be suspected in elderly patients with a sudden onset of severe osteopenia. Both discrete lytic lesions and diffuse osteopenia are suggestive factors in such circumstances.

The clinical management of osteopenia aims to identify and eliminate the cause of the condition. Nonspecific skeletal health promotion seeks to minimize bone loss and maintain skeletal competence; specific disease prevention and therapy attempt to eliminate, contain, or neutralize the cause of the disease. When the precise etiology and pathogenesis of a condition are poorly defined, therapy may be based on general principles rather than on intervention strategies that specifically target the disease. When osteopenia is attributable to lack of active vitamin D, specific intervention requires prescription of calcitriol. When osteopenia is the result of steroid therapy, normalization of steroid hormone levels, although the logical initial step in management of the skeletal problem, may be contraindicated in view of the underlying condition for which steroids have been prescribed. Intervention based on causal diagnosis requires that the primary cause of the condition be identified and appropriately addressed. In theory, causal diagnosis leads to specific therapy; in practice, this is only possible when causal diagnosis is accompanied by an adequate understanding of the underlying pathophysiological mechanisms. In osteoporosis, the cause of osteopenia is obscure.

Osteoporosis, the most prevalent clinical presentation of osteopenia, is recognized as a major health problem in developed countries. The financial cost of the acute care of this condition in the United States is estimated to be between seven and ten billion dollars annually (3). Because osteoporosis is a condition whose increasing prevalence parallels aging, and western demographic trends are toward an aging population, an intervention approach that attempts to shift the whole osteoporosis incidence curve is a desirable strategy from the perspective of population health promotion. Such a "population-based" strategy seeks to control the etiological determinants of a condition in the community as a whole. When the determinants of a condition are scientifically proven, intervention by this strategy offers a radical approach in which community behavior change can achieve a profound effect on the health status of the population. Ironically, the prevention paradox ensures that a preventive measure that brings much benefit to the population offers little to each participating individual (4). Truncation of the osteoporosis distribution curve, while increasing the overall cost of prevention, nonetheless achieves a favorable benefit–risk ratio for a few. Although "high-risk" intervention strategies are palliative in the sense that no attempt is made to address the underlying problem, osteoporosis disease prevention by the early recognition of vulnerable persons is a salutary activity in primary practice. In contrast to the population-based strategy, the high-risk strategy holds significant gains for the individual.

Currently, the major problem in attempting skeletal health promotion to avoid osteoporosis is related to uncertainty about the relationship of postulated risk factors to the pathogenesis of the condition. When pathogenesis is uncertain, a major guide to specifically targeting intervention strategies is lacking. Additional problems that potentially impede the early diagnosis of covert disease are the unavailability of adequately sensitive, specific, noninvasive, and inexpensive screening techniques. Conflicting research findings contribute to management dilemmas and impede the formulation of therapeutically effective intervention strategies.

Before diagnostic and intervention strategies can be selected, it is necessary to consider some aspects of bone physiology.

BONE METABOLISM

Skeletal health depends ultimately on an appropriate nutritional environment and weight bearing exercise. The chemical structure of bone consists of mineral (45%), water (25%), protein (20%), and fat (10%). The mineral content of bone is predominantly calcium and phosphate; in fact, 99% of calcium and 85% of phosphate in the body are in the skeleton. The major protein of bone is collagen. Bone formation involves the deposition of calcium phosphate as hydroxyapatite and amorphous calcium phosphate in a protein matrix. The protein matrix of bone is called osteoid and is produced by mononuclear osteoblasts. Mineralization of osteoid is essential for the normal development of bone. Osteoid osteocytes regulate the mineralization of the mature matrix. The rate of mineralization of osteoid depends on the concentration of calcium and phosphate ions in plasma and extracellular fluid. When the concentration of these ions in the extracellular fluid falls below a critical level, mineralization of bone is halted and demineralization may be initiated to restore a physiologically acceptable calcium ion concentration in the extracellular fluid. Osteocytes are capable of resorbing the inner areas of bone, and osteoclasts resorb the surface bone. Despite its apparent anatomical inertia, bone is a physiologically dynamic tissue. Bone remodeling takes about four months. On average, 8% of the skeleton is renewed annually through the organized process of bone resorption and bone formation. Trabecular bone, because of its high surface-to-volume ratio, has a high turnover rate—about eight times that of cortical bone. After peak bone mass has been achieved, slow inexorable bone loss is thought to occur at a rate of about 0.3–0.5% annually in cortical bone and about 1.2% annually in trabecular bone (5). Cortical bone comprises 80–85% of the total bone in the body; trabecular bone is mainly found in the vertebrae (lumbar vertebrae are 66% trabecular bone), the neck of the femur (25%), and the distal end of the radius (25%).

The basic structure of bone mineral is hydroxyapatite: $Ca_{10}(PO_4)_6(OH)_2$. Hydroxyapatite is initially in an amorphous form and later assumes a crystalline structure. There is a constant flux of calcium and phosphate between blood, tissue, and bone reserve. Vitamin D plays an important role in the regulation of body calcium and phosphate levels. The exact structure of hydroxyapatite varies with age, site, and chemical composition of the bathing fluid. With age the amorphous hydroxyapatite decreases and the crystalline form increases; bone becomes harder. Traces of zinc, sodium, magnesium, and potassium may be found in bone crystal. As bone is a dynamic living tissue, ion exchange constantly occurs between the hydration shell surface and the interior of the apatite crystal. Substitution of ions in the hydroxyapatite crystal is not an irreversible process. Fluoride can replace the hydroxyl ion; calcium can be replaced by sodium, magnesium, lead, strontium, barium, or radium. Fluoride substitution of the hydroxyl ion reduces the solubility of the mineral component of the crystalline structure. In addition to calcium, phosphorus, and fluoride, which have particular significance for bone mineralization, silicon and zinc are considered important for bone health. Silicon is probably important as part of the intercellular ground substance in which collagen is embedded, and zinc is a cofactor in nucleic acid and protein synthesis. Toxic elements such as aluminum, lead, cadmium, and the radioactive elements plutonium and strontium may be deposited in bone. Deposition of radioactive elements adjacent to bone marrow may result in hemopoietic disturbances. Iron stores are found in the bone marrow core of long bones and ribs.

Collagen, the major protein of bone, is synthesized by osteoblasts. In addition to requiring amino acids, particularly glycine, proline, and hydroxyproline, collagen requires vitamin C and copper for its synthesis. Vitamin C is necessary for the hydroxylation of certain amino acids in procollagen synthesis; hydroxyproline and hydroxylysine are involved in the characteristic coiling of procollagen. Lysyl oxidase, a copper-activated enzyme, is involved in disulfide cross-linking of the procollagen strands during collagen synthesis. Mineralized bone not only serves an important structural support function, it is the major calcium storage organ in the body. Any physiologically significant modification in extracellular calcium ion concentration is homeostatically corrected. Bone serves as a calcium reserve for such homeostasis. Calcium plays a vital role in neuromuscular transmission, membrane excitability, and muscle contractility; it also mediates hormone action, participates in blood coagulation, and acts as a cofactor in certain enzyme reactions. Hormones regulating the mechanism responsible for calcium include active vitamin D (calcitriol), parathyroid hormone (PTH), and calcitonin. Calcium homeostasis is achieved by modifying calcium absorption, urinary calcium excretion, and bone resorption.

Calcitonin secretion increases in response to an increase in serum calcium and decreases in response to a drop in serum calcium. Females have lower circulating levels of calcitonin than males, and these levels may further decrease with age. Calcitonin probably works at the level of the intestine, kidney, and bone through the adenyl cyclase system and production of cyclic AMP. The effects of calcitonin are:

- To inhibit osteoclastic activity, thereby decreasing bone resorption without impairing new bone formation. Osteoclasts are the cells responsible for the resorption component of bone remodeling.
- To increase urinary excretion of calcium, phosphate, sodium, and potassium.

Calcitonin has also been reported to promote the healing of wounds and fractures.

Although the major biological action of calcitonin is to inhibit osteoclasts with the consequent inhibition of bone resorption, there is currently no convincing evidence to suggest that osteoporosis may be the result of calcitonin deficiency. As calcitonin impedes bone resorption but does not influence bone formation, its potential usefulness lies in osteoporosis prevention rather than skeletal health promotion (6). This hormone helps to maintain existing bone mass. In contrast to vitamin D and its metabolites, which appear to be unhelpful in the treatment of postmenopausal osteoporosis, pharmacological doses of calcitonin do appear to restrain bone loss. Once a satisfactory mode of administration (probably intranasal to avoid gastric digestion) and an economically viable means of production (possibly bacterial) have been elucidated, the prospects for calcitonin in preventing postmenopausal osteoporosis appear promising (7). Calcitonin may significantly alter current intervention strategies, should suggestions prove correct that estrogen's protective effect on the skeleton of postmenopausal women is mediated by calcitonin production. A two-year randomized pilot study found that in doses exceeding 50 IU a week, calcitonin was as effective as estradiol in reducing vertebral bone loss (8).

Like calcitonin, PTH secretion is also largely regulated by the serum concentration of ionized calcium. There is, however, an inverse relationship between ionized serum calcium concentration and PTH secretion: a fall in serum calcium stimulates PTH output; an increase in serum calcium inhibits PTH output. Magnesium ion concentration and vitamin D metabolites may also influence PTH secretion. PTH, mediated by cyclic AMP, acts on bone and the kidney. The major effects of PTH secretion are:

- To increase calcium and reduce phosphate reabsorption in the kidney.
- To enhance calcium mobilization from the skeleton by stimulating osteoclastic activity in bone. The action of PTH on bone is concentration-dependent; target organ receptor modification in response to PTH may partially explain some of the complexity associated with this hormone's action.
- To potentiate vitamin D in its intestinal absorption and renal conservation of calcium.

Hyperparathyroidism, a condition associated with excess PTH secretion, presents clinically with symptomatology attributable to hypercalcemia, renal and urinary tract damage, and skeletal involvement. The major manifestations of hypercalcemia are thirst, nausea, vomiting, fatigue, and depression. Renal manifestations include polyuria, renal calculi, and nephrocalcinosis. Skeletal symptoms range from bone pain to pathological fractures. Features of osteomalacia (inadequate mineralization), osteoporosis (reduced bone density), and osteosclerosis (increased bone density) may be found.

The last of the major hormones influencing bone homeostasis is vitamin D. There are different forms of active vitamin D; vitamin D_2 is of plant origin and vitamin D_3 is of animal origin. Cholecalciferol, vitamin D_3, is produced by irradiating the 7-dehydrocholesterol synthesized in the liver with ultraviolet light. In humans this reaction takes place in the stratum corneum of the skin; it only occurs when the skin is exposed to direct sunlight. Other, less important, sources of cholecalciferol in humans are animal fats such as cod liver oil, salmon, sardines, eggs, butter, cheese, and liver. Cholecalciferol, derived from the skin and from intestinal absorption, is converted to 25-hydroxycholecalciferol in the liver, where it is stored. Measurement of plasma 25-hydroxycholecalciferol may be used as an indication of body vitamin D stores. The active form of vitamin D_3 is 1,25-dihydroxycholecalciferol (calcitriol). The final hydroxylation takes place in the kidney. After double hydroxylation the active form of vitamin D_3 so formed is ten times more potent than its precursor. The overall effect of calcitriol to increase plasma calcium levels is achieved by modifying calcium and phosphate metabolism by:

- Promoting intestinal calcium absorption. Active vitamin D_3 induces synthesis of a specific calcium binding protein in epithelial cells.
- Mobilizing calcium stored in bone. This action is potentiated by PTH.
- Facilitating phosphate absorption in the intestine by means of a calcium-independent mechanism.
- Enhancing urinary calcium reabsorption.

Clinically, vitamin D deficiency presents as rickets in children and as osteomalacia in adults. Rickets is characterized by shortened bones, which may be bowed if weight bearing has been permitted, enlarged joints, and a rachitic rosary representing swelling of the osteochondral junctions of the rib cage. The child is listless, flabby, and sweaty. Serum alkaline phosphatase is elevated. The adult form of vitamin D deficiency, osteomalacia, must be differentiated from osteoporosis, the other major disorder of adult bone metabolism. Osteomalacia is a condition in which undermineralization of bone gives rise to generalized musculoskeletal symptoms with persistent pain and muscle weakness. Osteoporosis is a decrease in bone density seen on plain radiographs as osteopenia; symptoms are localized to fracture sites. In osteomalacia the calcification rate is decreased, osteoid accumulates, and serum alkaline phosphatase is elevated. Serum calcium and phosphate levels may be low. In osteoporosis bone is normally mineralized, the rate of bone resorption being increased; biochemical indices are slightly raised or within normal limits. In osteomalacia, the defect is attributable to inadequate mineralization of bone; in osteoporosis, the calcification rate is normal and the reduction in bone mass may well be a reflection of accelerated resorption. Etiologically, osteomalacia is attributable to a deficiency of active vitamin D_3; the primary condition responds dramatically to adequate active vitamin D_3 administration. The etiology of osteoporosis is uncertain; when the pathogenesis is unknown treatment is based on innovative rather than selectively proven intervention.

Associated with the homeostatic system in which maintenance of an equable ionized serum calcium level takes precedence over skeletal health is a sophisticated dynamic bone remodeling system. The maintenance of a mechanically competent skeleton depends on the ability of the bone to constantly adapt its gross and microscopic organization, to correct microdamage by continuous bone tissue renewal, and to maintain a bone volume commensurate with "usual" mechanical demands. Jaworski (9) suggests that effective bone remodeling requires mechanical loading and is influenced by permissive and modulatory factors. Permissive factors include oxygen; nutrients such as calcium, phosphorus, and amino acids; and hormones such as cortisol, insulin, and thyroxin. Modulatory factors include PTH, which links calcium and phosphate ion homeostasis to the bone remodeling system, and secondary sex hormones, notably estrogen, which regulates bone balance using the skeleton as a mineral storage depot. Alterations in mechanical load and in permissive or modulatory factors are postulated to alter the dynamic balance between osteoblastic and osteoclastic cellular activity—between bone formation and resorption. The relationship between these variables provides the basis for models used to explain the etiology and pathogenesis of osteoporosis. Two such models postulate calcium and/or estrogen deficiency.

ETIOLOGY AND PATHOGENESIS OF OSTEOPOROSIS

Despite extensive study, the precise etiology and pathogenesis of primary osteoporosis remain obscure. Postulates range from suggestions that the key pathogenic event is bone cell senescence to theories proposing that osteoporosis should be viewed as a deficiency disease.

Calcium Deficiency Hypothesis

Even if the calcium deficiency hypothesis is correct in attributing osteoporosis to an absolute or relative lack of calcium, the pathogenesis of this condition remains obscure. The mechanism whereby a calcium deficiency results in relentless bone demineralization with subsequent development of porous, light, weak, easily fractured bones has yet to be clarified. Postulates range from dietary calcium deficiency causing secondary hyperparathyroidism leading to cortical bone loss, to increased postmenopausal calcium requirements attributable to impaired activation of vitamin D (and calcium absorption) in the absence of adequate estrogen levels. In reality, many people do not meet the recommended daily intake of calcium, and circulating levels of active vitamin D_3 are reduced in elderly patients.

Despite uncertainty about the pathogenesis of calcium deficiency in the development of osteoporosis, the calcium deficiency model for osteoporosis "probably represent[s] the most powerful conceptual model for the understanding of this condition at present" (10). This model is based on the premise that blood calcium homeostasis takes precedence over skeletal preservation. Since 99% of the body's calcium is located within the skeleton, calcium homeostasis certainly deserves consideration as a determinant of bone density. Nordin (11) perhaps puts the calcium question in proper perspective when he suggests that calcium deficiency, in the sense of dietary inadequacy, is no more the single or major cause of osteoporosis than is a high salt intake the sole or main cause of hypertension, or high dietary cholesterol the sole or main cause of coronary artery disease. Any suggestion that osteoporosis is a unifactorial disease attributable to dietary calcium deficiency is a gross and inaccurate oversimplification. Nordin furthermore points out that the relationship of postmenopausal bone loss to estrogen deprivation no more precludes the calcium deficiency model than does the frequent relationship of iron deficiency anemia to blood loss negate the iron deficiency model of anemia. Blood loss may be regarded as resulting in an increased iron requirement; estrogen deprivation may be regarded as causing an increased calcium requirement. Nordin and associates (10) explain the role of calcium deficiency in the pathogenesis of postmenopausal osteoporosis by postulating that the primary event in postmenopausal osteoporosis is an increased serum calcium, which triggers increased calciuria and an increased calcium requirement. In this regard the authors note that the subset of the postmenopausal population that develops spinal fractures shows additional risk factors such as an inability to absorb calcium and reduced adrenal androgen production. In any event, evidence does suggest that menopausal bone loss is attributable to bone resorption.

In determining calcium status, two factors must be taken into account.

1. Plasma calcium homeostasis takes precedence over preservation of calcium within the skeleton. The plasma regulation of calcium levels is mediated by the parathyroid glands, which mobilize calcium from the intestine (through production of 1,25-dihydroxy vitamin D), the renal tubules, and the skeleton. PTH mobilizes calcium from the skeleton by osteoclastic resorption; this results in a loss of bone density. With longstanding calcium shortage, PTH induces monocytic precursor cells to transform to osteoclasts. Calcitriol probably acts on the precursor phase of osteoclasts.

2. The plasma calcium level does not provide a guide to a person's calcium nutritional status. Plasma calcium estimations cannot discriminate between individuals in positive and negative calcium balance. Assays of blood PTH levels are themselves too imprecise and regulators of active vitamin D_3 are too numerous (calcium intake, phosphorus intake, growth hormone, prolactin, etc.) to provide adequate markers of calcium deficiency. It has therefore been proposed that only calcium balance studies provide an adequate measure of calcium nutritional requirement.

Calcium deficiency is a particularly attractive theory when one considers that normalization of any potential reduction in serum calcium can theoretically be achieved either by increasing calcium absorption or by reducing urinary excretion (12).

Calcium Homeostasis

Calcium homeostasis implies the control of serum calcium within clearly defined limits. Any reduction in serum calcium is rapidly modified by PTH secretion, which corrects the calcium balance by increasing calcium absorption from the bowel, enhancing calcium reabsorption from the urine, and withdrawing calcium from bone, the calcium store. Dietary minimization of calcium deficiency can be achieved by providing adequate dietary calcium, maximizing calcium absorption, and minimizing calcium excretion.

Adequacy of Dietary Calcium

In 1979 the recommended daily allowance (RDA) of calcium for Australian adults was 40–800 mg; this was increased in pregnancy and lactation to 900–1,300 mg (13). In July 1985 these RDA levels were revised, and the increased Australian RDA levels for calcium—1,000 mg for women over 54 years and 800 mg for women under 54 years and for all men—were passed at the 100th session of the National Health and Medical Research Council. The RDA for American women is marginally lower. Despite the international similarity of recommendations, balance studies suggest that RDA levels for calcium are inadequate (14).

Certain balance studies have suggested that premenopausal women have a calcium requirement of 1,000 mg/day, and estrogen-deprived women require 1,500 mg/day (15). A more recent study shows that at dietary calcium levels of 800 mg/day, half of the estrogen-deprived women studied were in negative calcium balance; nearly one quarter remained in negative calcium balance at daily calcium levels of 1.5 g (16). Furthermore, balance studies, when used as a basis for determining RDAs, are usually derived from studies on young adults with zero balance as the hallmark of nutritional adequacy. A calcium intake that sustains a zero balance in an adult who has attained peak bone mass may not provide an optimal environment for maximization of bone mass in a teenager or younger adult yet to achieve peak bone mass. If it is assumed that a 70 kg individual contains approximately 1,200 g of calcium, the net calcium gain from birth to peak bone mass should be of the order of 1,175 g. If peak bone mass is reached at 25 years of age, this implies an accumulation of almost 130 mg of calcium each day over a 25 year period. While a daily dietary calcium intake of 800 mg may questionably permit calcium balance, it does not allow for the positive calcium balance required for peak bone mass if one assumes a 30% dietary calcium absorption and an obligatory daily calcium excretion of 250 mg. In the past, RDA levels for calcium were found inadequate and required revision; it is possible that the current recommendations may be subject to a similar escalation at some time in the future.

While the adequacy of calcium RDAs may serve as a hypothetical source of concern for skeletal health, actual calcium dietary intake constitutes a real dilemma. An estimate of the calcium intake of American women suggests that their daily dietary intake of calcium is about 600 mg; postmenopausal women, whose calcium need is increased, appear to further reduce their daily dietary calcium to about 480 mg (17). An Australian study found that the calcium consumption of 29% of participating postmenopausal women was less than 500 mg/day, with 47% of the total study population ingesting less than 740 mg/day (18). From their analysis of the literature, Angus and Eisman (19) concluded that the mean calcium intake of postmenopausal women in Australia, Britain, and the United States is probably of the order of 700–750 mg/day. Mean calcium intake levels vary from a low of 500 mg/day to a high of 900 mg/day, depending on the dietary assessment technique. Techniques that are easier to implement, e.g., 24 hour recall, are recognized as less precise than the more demanding multiple-day weighted food diary.

This survey of the literature also found that more than 60% of American and Australian premenopausal women did not meet their country's RDA for calcium; this figure increased to 75% and over for postmenopausal women of both countries. Based on current RDA levels, many women consume diets that are calcium deficient.

Calcium Absorption

In the Australian or American diet, the percentage absorption of calcium is thought to be about 20%. Serum calcium is maintained within finely controlled limits; low dietary intake is associated with enhanced absorption (about 30%), while absorption from diets high in calcium is proportionately less. Adaptation to dietary composition is therefore possible—within limits. Calcium is found to some extent in vegetables, is plentiful in sesame seeds, and is abundant in canned fish when the bones are eaten, but the main sources of calcium in the western diet are milk and milk products, e.g., cheese (20). The latter are also sources of animal fats. Dietary recommendations of various health departments routinely include a reduction in fat consumption. As dairy products are a rich source of calcium, and absorption of dietary calcium is enhanced by lactose, fat-reduced dairy products deserve recognition as an important dietary source of calcium. Low-fat, high-energy diets permit up to 1.4 liters of skim milk each day, and bone remodeling appears to progress more satisfactorily with milk than with calcium carbonate (21). As a general rule of thumb, at least 60% of daily calcium intake should come from dairy products. An American study estimates that 75% of dietary calcium is derived from milk and dairy products; an Australian study estimates 59% (19). In either case, it is unlikely that the individual patient is meeting the RDA for calcium if milk and dairy products are excluded from the diet. Patients on low-energy diets also run an increased risk of calcium deficiency. This deficiency may be further aggravated by a diet rich in complex carbohydrates.

Dietary fiber impairs calcium absorption (22). Cellulose, phytates, and oxalates all impair calcium absorption. Spinach, because of its high oxalate content, is a rich but biologically unavailable source of calcium. Data on the effect of dietary fiber on calcium balance are limited and somewhat contradictory (23), but it has been suggested that each 18 g of fiber ingested raises the calcium requirement by 100 mg. The major source of calcium loss is fecal excretion; calcium unavailable for absorption is excreted in the feces.

Reduced calcium absorption has also been postulated as a mechanism whereby alcohol enhances the risk of osteoporosis. Other possible mechanisms include direct suppression of osteoblast function and increased circulating cortisol levels. Cortisol levels are raised in heavy drinkers owing to direct stimulation of the hypothalamus or impaired hepatic clearance.

Calcium Excretion

Calcium absorption and excretion respond to changes in calcium levels within 24 hours. Calcium absorption must equal calcium loss to achieve calcium balance. Absorbed calcium is excreted in urine and sweat. Obligatory calcium losses are accepted as being about 250 mg/day. For calcium balance this assumes an average absorption of 30% on an RDA intake of 800 mg/day. Attempts to preserve calcium balance may be facilitated by minimizing calcium excretion. Obligatory calcium losses occur in the urine, feces, and perspiration.

1. Calciuria. Obligatory excretion of calcium in the urine, on a standard western diet rich in protein and salt, is about 120 mg/day. Given calcium physiology, a daily intake of 150 mg is required to counter this loss. About 1–2% of filtered calcium has to be excreted from the tubular filtrate. Some 60% of plasma calcium undergoes ultrafiltration at the glomeruli; calcium is reabsorbed along the nephron by diverse mechanisms. In the proximal convoluted tubule, outward calcium transport is determined by the reabsorption of salt and water; 10% of the filtered load arrives at the distal convoluted tubule. In the distal tubule calcium reabsorption is by an active transport mechanism stimulated by PTH and dependent on calcitriol. Absorption of calcium by the distal tubule is impaired by hypercalcemia and metabolic acidosis.

2. Fecal losses. Obligatory endogenous calcium loss in the feces is of the order of 200 mg/day. Some physiologists estimate obligatory fecal losses to be 2–5 mmol/day. Endogenous fecal calcium is derived from intestinal secretions and desquamated cells. Intakes above 200 mg undergo rapid intestinal absorption by active transport mechanisms regulated by calcitriol; once this active transport mechanism is saturated, further absorption is by diffusion. Calcitriol acts as a steroid hormone, both inducing intestinal synthesis of proteins connected with the transcellular transport of calcium and influencing the permeability of the brush border to calcium. Most calcium absorption is by means of active absorption in the proximal small intestine.

3. Losses in perspiration. Small amounts of calcium are lost in sweat (0.5 mmol/day).

Sodium intake has been shown to modify calcium excretion. An increase in dietary salt intake is associated with an increase in the obligatory excretion of calcium (24,25); sodium and calcium excretion have been shown to have a linear relationship. One additional teaspoon of salt causes an approximate increase of 100 mmol/day of urinary sodium, and this is associated with an increase of about 1.3 mmol/day of urinary calcium (26). A minor alteration (0.02 mg/100 ml) of calcium in the glomerular filtrate may double or halve urinary calcium excretion. The equilibration of calcium absorption and urinary excretion is only normally achieved when the net absorbed calcium equals or exceeds 150 mg; absorption of less than 150 mg of calcium per day results in a negative calcium balance due to obligatory excretion of this mineral. Taste adaptation to salt restriction does occur after about two months of salt restriction (27). Diets with salt levels of less than 0.3 g/day are unpalatable—not even sodium restriction in the management of hypertension requires such stringent restriction. The mechanism underlying sodium-induced calciuria is postulated to be mediated by PTH activation of vitamin D (28).

Like high-sodium diets, high-protein diets are believed to increase the obligatory urinary excretion of calcium (29). High-protein diets lead to an increased excretion of acidic sulfuric products with a concomitant reduction in the reabsorption of urinary calcium. The extent of the calciuria associated with high dietary protein may be minimized by consumption of protein-rich foods that are also rich in phosphorus (30), such as meat and dairy products. Even if dietary calcium is plentiful, the net effect of very high-protein diets (142 g protein per day) is a persistently negative calcium balance (31). It is worth noting that calcium absorption in normal male subjects increased with increased dietary protein when the daily dietary calcium concentration was 800–1,400 mg (32). No increased calcium absorption was noted at dietary calcium levels of 500 mg. Ingestion of small amounts of calcium bicarbonate may increase retention in protein-induced calciuria (33). It is also worth remembering that the urinary loss of calcium resulting from a dietary increase of one teaspoon of salt is greater than that following a 50% increase in protein consumption.

Less well established is the calciuric effect of caffeine. One study has linked caffeine intake with a mildly increased calciuria, but another study has failed to show any correlation between current caffeine intake and distal radial bone density. No data relating caffeine intake to vertebral bone maintenance are reported in the literature.

Links between dietary calcium, calciuria, and bone mass are complex. Correlational studies show that bone density is a significant inverse function of urinary calcium and age, and a significant positive function of dietary calcium. The rate of bone loss is a function of initial bone density, but when this and urinary calcium are included in the calculation, dietary calcium is no longer significant whereas urinary calcium remains statistically important. Hydroxyproline is strongly related to urinary calcium and, when this is allowed for, inversely related to dietary calcium. The higher the urinary calcium and the lower the calcium intake, the higher the hydroxyproline and the lower the bone density; nonetheless, the role of dietary calcium cannot be directly demonstrated in the rate of bone loss.

The relationship between phosphorus and bone mass maintenance is equally confusing (34). The ratio of dietary calcium to phosphate (Ca:P) is believed to influence calcium metabolism. Most diets have a Ca:P ratio of between 1:2 and 1:4. Animals require a 2:1 ratio. Humans appear able to adapt to the current dietary ratio provided their vitamin D status is adequate. While the ideal Ca:P ratio is 1:1, no ill effects are associated with the current dietary ratio provided it does not exceed 3:1 or 1:3 (35). Infants should not have a dietary Ca:P ratio of less than 1.5:1. Human milk has a Ca:P ratio of 1.6:1; in cow's milk the ratio is 1:1. Dietary phosphorus in the range of 160–2,270 mg/day probably has no effect on calcium balance. While some authors consider it unlikely that typical modern dietary Ca:P ratios have any effect on calcium balance and bone maintenance, others postulate an excessive phosphorus ingestion in contemporary processed foods. Such excess phosphorus is hypothesized to initiate a state of tertiary hyperparathyroidism in which alveolar bone recession proceeds to full-blown osteoporosis.

Phosphorus and Magnesium Balance

Phosphorus balance is clearly influenced by calcium metabolism. Hormonal control is exerted at the level of both bone and kidney. Calcitriol stimulates phosphate uptake from the intestine and increases mobilization of phosphate from bone. At the proximal renal tubule, reabsorption of calcium and phosphate is differentially controlled. It is here that the critical balance of calcium relative to phosphate is determined. When serum calcium is low and serum phosphate is high, the vitamin D–parathyroid hormone system increases calcium and decreases phosphate reabsorption; when serum calcium is high and serum phosphate is low, calcitonin release increases overall calcium losses and increases phosphate retention.

Phosphate deficiency is unusual given the contemporary diet; it can occur in persons with renal disease or with extensive use of drugs, e.g., an aluminum hydroxide antacid mixture. Phosphate excess manifests clinically with neonatal hypocalcemic tetany. This is encountered in babies fed undiluted cow's milk. Cow's milk has six times the phosphate concentration of human milk. The baby has twitches and spasms. The therapy is calcium replacement to normalize the serum Ca:P ratio.

Magnesium is another mineral that may influence calcium metabolism. Magnesium is an important intracellular ion. The total body content of magnesium is 25 g, half of which is stored in bone. Body magnesium levels are regulated by intestinal absorption and renal tubular reabsorption, i.e., urinary excretion. Dietary absorption of magnesium is 30–40%; this percentage increases when plasma magnesium is low and decreases with magnesium sulfate ingestion. The RDA for magnesium is 350 mg for men and 300 mg for women. Green vegetables are a good source; magnesium is the central atom of chlorophyll. It should be noted that vegetables, e.g., lettuce, can flourish in magnesium-depleted soils and be a poor dietary source of this mineral.

Magnesium deficiency may occur in severe malabsorption, in diarrhea, in renal dysfunction, and in chronic alcoholism with malnutrition. The clinical presentation of magnesium deficiency is similar to that of calcium deficiency: anorexia, nausea, apathy, and muscle tremor. In severe cases coma and convulsions may occur. The similarity of the clinical presentations of calcium and magnesium deficiency is explained by the apparent requirement for magnesium as a cofactor in calcium mobilization. In humans, magnesium deficiency produces a drop in serum calcium.

Calcium Intake: A Summary

It is logical to suppose that in order to maintain bone calcium content, obligatory calcium losses are replenished by dietary calcium. Low calcium intake does appear to be undesirable for skeletal health (36) and dietary calcium deficiency has clearly been linked to osteoporosis in animals (37), but the etiological and therapeutic relationship between calcium consumption and osteoporosis in humans remains controversial (38–40).

> The opinion that calcium intake is not clearly related to bone mass or to osteoporosis is well-entrenched in the nutrition literature. . . . [T]his position rests on

the poor correlation between habitual calcium intake and the incidence of osteoporosis in some non-Western countries, and on the demonstration that young men can achieve calcium balance on low intakes. However, such data neglect a variety of factors determining calcium nutriture (30).

One of the most consistent observations, which cannot be ignored, is the inadequate calcium intake of women in the second and third decade of life and the negative calcium balance that characterizes the perimenopausal state (41).

More precisely, it has been postulated that there may be a calcium intake threshold below which calcium supplementation may be beneficial in reducing spinal mineral loss; a clinical trial showed that women with a calcium intake of less then 405 mg/day lost significantly more spine density than those whose calcium intake exceeded 777 mg/day (42). Angus and Eisman (19) quote three different studies that support a strong link between calcium intake in early life with peak bone mass: "In spite of the fact that the role of dietary calcium in age-related osteoporosis remains controversial, an adequate calcium intake throughout life is a sensible approach." It would certainly be most unwise for teenage girls to consume less than the RDA of calcium at a time when their bone mass has not yet peaked.

Estrogen Deprivation

Estrogen deprivation at menopause is thought to result in an uncoupling or imbalance between bone resorption and bone formation. Greater loss of endosteal bone than synthesis of periosteal bone results in the formation of bones that are wider and have a thinner cortex. During the early postmenopausal period there is a primary increase in osteoclastic bone resorption, which lasts for about ten years and affects both cortical and trabecular bone. In older age groups a primary decrease in bone formation is attributed to either a decrease in recruitment of new osteoblasts to repair bone cavities (resorption bays) or a general decrease in the activity of osteoblasts that have accumulated at the resorption site. Hypotheses describing the mechanism underlying the action of estrogen include:

- The presence of estrogen receptors on osteoblasts.
- Estrogen antagonism of PTH, either at the point of release of PTH or at the cellular level of PTH activity.
- Estrogen-stimulated release of calcitonin. The mechanism of this stimulation is unknown but is postulated to be either a direct effect of estrogen on calcitonin-secreting cells of the thyroid or mediated through estrogen-enhanced intestinal calcium absorption. Calcitonin responds to hypercalcemia by suppressing bone resorption and renal tubular reabsorption of calcium, thereby lowering blood calcium concentrations. Although the mechanism of estrogen-stimulated release of calcitonin may be unclear, calcitonin itself is clearly associated with a reduced risk of osteoporosis. Circulating calcitonin levels are higher in blacks than in whites, in males than in females, and in the young than in the elderly. Animal experimentation supports the notion that calcitonin has a protective effect against bone

loss; total thyroidectomy leads to decreased bone mineral content after some years; and plasma calcitonin levels provoked by calcium challenge are lower in patients with vertebral fractures than in controls (43).
- Estrogen-enhanced conversion of 25-hydroxyvitamin D to calcitriol by renal tubular cells.
- Possible estrogen-influenced production of local factors that influence bone enzymes. Local regulators postulated to influence bone growth include monokines and lymphokines derived from blood cells, bone- and cartilage-inducing factors derived from bone matrix, and growth factors derived from cartilage and bone.

The most rapid period of bone loss associated with estrogen deprivation occurs during the first five to ten years after menopause. Cortical bone loss is initiated at about 40 years of age and progresses at a rate of 0.3–0.5% per annum in both sexes. In the five to ten years following menopause, cortical bone loss accelerates to 2–3% per annum and then gradually returns to the preclimacteric rate of loss. In women, trabecular bone loss begins at about 30 years of age, progresses at a rate of 0.6% per annum, and then accelerates to 2.4% for two or three years following menopause. The effects of estrogen deprivation appear to be confined to the perimenopausal period, and estrogen supplementation is not recommended for women who have had no estrogen replacement for the first ten to 15 years postmenopausally. Estrogen stabilizes but does not restore bone mass. The first six to 24 months of estrogen replacement are accompanied by an increased bone density; with continued therapy bone mass stabilizes, and in time there is a return to the gradual bone loss associated with aging. Estrogen appears to disallow accelerated bone resorption (44).

Certain factors that have been correlated with the risk of developing osteoporosis can be explained in terms of the estrogen deprivation hypothesis. Obesity is believed to provide some protection against osteoporosis; the peripheral conversion of androgens to estrogens in adipose tissue may contribute to this protective effect. Skeletal stress associated with weight bearing is an alternative explanation for this finding. Smoking, on the other hand, is thought to increase the risk of osteoporosis. Some postulated mechanisms for this association are a reduced bioavailability of estrogen in smokers, interference with the microcirculation in bone, and appetite suppression, which facilitates maintenance of a reduced body weight.

Some recent studies dispute any link between smoking and bone mass. Using single photon absorptiometry to compare bone mass in smokers and nonsmokers, no difference was found at either the distal or the middle portion of the radius; the researchers concluded that smoking has no direct effect on the appendicular skeleton (45). This study did not control for confounding variables such as body weight and diet. Another study assessing single photon absorptiometry as a screening tool in osteoporosis found that middle and distal radial bone mass correlated significantly with hereditary factors but not with calcium intake, activity level, or smoking (46). Clearly, not only the etiology and pathogenesis influencing bone mass but also the screening techniques used to assess bone mass require further investigation.

Summary

It would be erroneous to assume a one cause–one disease approach to osteoporosis. Postmenopausal women within one year of menopause appear to undergo an increase in obligatory calcium loss. The calcium deficiency hypothesis would support the suggestion that osteoporosis is a disease of calcium deficiency resulting not from absolute dietary deficiency, but from impaired calcium absorption. Calcium absorption may be impaired in postmenopausal women. A prospective clinical trial on 171 postmenopausal women suggested that a calcium absorption defect may be present in a significant number of apparently normal women (16). Calcium and estrogen deficiency may be two of a possible range of variables contributing to the multifactorial etiology of osteoporosis.

ASSESSMENT OF OSTEOPOROSIS

A precise knowledge of pathogenesis facilitates a focused intervention, and an accurate patient assessment targets individuals requiring intervention. The early recognition of patients with a propensity for osteoporosis depends largely on the ability to detect either radiological modifications in bone mass/density or chemical changes suggestive of an imbalance in bone remodeling. The poor correlation between bone mass and fracture severity suggests that factors in addition to bone mass are important in determining whether a bone will fracture. These variables have not been adequately defined and are therefore not helpful as diagnostic criteria. Height loss, which correlates well with the fracture index, is an overt sign of osteoporosis and a late sign in the pathogenesis of the condition; it too correlates, but weakly, with bone mass measurements. Despite these limitations, assessment of bone mass remains a major means of ascertaining the individual's risk of overt osteoporosis.

The least invasive and costly of bone density assessments is the plain x-ray, but it is of little use as a diagnostic tool for elderly osteoporosis intervention. Loss of 40–50% of bone mass occurs before this is detectable on plain x-ray (47,48). More sensitive measurement of bone mass may be achieved using techniques such as single and/or dual photon absorptiometry and quantitative computed tomography. These more sophisticated techniques are better predictors of total body calcium.

Single photon absorptiometry, usually of the forearm and occasionally of the calcaneus, has proved a useful means of assessing the bone density of the appendicular skeleton. Not only is this technique an acceptable predictor of future fractures at appendicular sites not scanned, it is also useful for follow-up therapy of osteopenia affecting the appendicular skeleton. Single photon absorptiometry is painless, is quick, and uses a small dose of radiation. Although accurately measuring forearm bone density, it fails to selectively predict femoral neck or vertebral bone mass. Events in forearm bones are not always reflected in vertebrae, and vice versa. Dual photon absorptiometry, although more expensive and time consuming than single photon absorptiometry, does provide a good index of the mineral content of the spine and hip (49). Dual photon absorptiometry, used to measure the bone density of the lumbar spine, femoral neck, midshaft femur, and/or whole skeleton, has also proved a helpful predictor of future fractures at sites not scanned and is a useful tool for monitoring the response of the axial skeleton to therapy. The presence of calcium in the aorta or of spinal arthritis increases the risk of false-negative results. As single photon absorptiometry is a poor predictor of changes in the axial skeleton, dual photon absorptiometry is a more useful means of assessing bones with higher concentrations of trabecular bone. The cortical:trabecular ratio of metacarpal bones is 99:1; that of the midradius is 95:5, the distal radius 75:25, and lumbar vertebrae about 40:60. Because osteopenia progresses most rapidly in the axial skeleton following postmenopausal estrogen deprivation, dual photon absorptiometry would seem to be the screening test of choice in high-risk patients. It is worth noting that single photon absorptiometry of the distal forearm correlates better with dual photon absorptiometry of the vertebrae than does the proximal forearm measurement (50). Both single and dual photon absorptiometry are useful research tools but unproven screening tools (51); nonetheless, some suggest that dual photon absorptiometry is more able to separate patients with osteoporosis from the normal population than is computed tomographic scanning (52).

Quantitative computed tomography of the spine requires a phantom standard and uses about ten times more radiation than photon measurements, and its reliability is adversely affected by fat in bone marrow. Nonetheless, axial sites can be scanned and tomographic cuts of vertebrae are possible using this technique (53). The value of quantitative computed tomography lies in its noninvasive nature, high precision, and high sensitivity for vertebral trabecular measurement (54). Because body weight, endogenous levels of sex steroids, renal calcium, and hydroxyproline levels are probably not related to individual bone mass in the lumbar spine, monoenergetic computed tomography may be particularly useful for establishing the trabecular bone mass of lumbar vertebrae (55).

Mineral densitometry uses computed tomography and/or nuclear medicine to measure the concentration of minerals in bone. The nuclear medicine technique cannot distinguish between the mineral content of the cortex (compact) and medulla (cancellous), but computed tomography can differentially measure the density of cancellous bone. Both techniques can be applied to the axial skeleton and proximal femur (56).

The normal range of forearm mineral density in young women is 370–570 mg/cc. Forearm mineral density peaks between the ages of about 20 and 25 years and remains relatively unchanged until menopause. Although forearm mineral density is highly correlated with vertebral density on a population basis, individual variations have been detected. Nonetheless, this is considered a practical procedure on which to base an objective osteoporosis intervention program. Nordin and Need (56) recommend that women have a midlife forearm densitometry measurement. Those with a bone density above the 75th percentile can be reassured; those with a bone density below the 25th percentile require further investigation and treatment; and those in between should undergo further assessment in a few years.

All techniques used to measure bone mass provide structural information. In a dynamic condition in which imbalances in bone turnover predicate disease, a more functional measure of bone activity would be desirable. No definitive techniques for measuring skeletal metabolism have been established. Like routine radiological tests, routine laboratory tests are not helpful. Total plasma or albumin corrected-calcium levels are at best a crude measure of hypo- or hypercalcemia and have no relevance in contemporary osteoporosis assessment. Measures used for monitoring skeletal health rely on the propensity of adult bone to undergo constant remodeling; bone-forming osteoblasts synthesize new bone to replace bone removed by bone-resorbing osteoclasts. In general, there is more osteoclastic than osteoblastic activity; one osteoclast is 225 times more efficient than one osteoblast. Bone mass may decrease as a result of either an absolute or a relative increase of bone resorption, or an absolute or a relative decrease in bone formation. Such bone formation/resorption imbalance may occur in an environment where overall bone remodeling is increased or reduced.

Osteocalcin or bone Gla protein is a specific product of osteoblasts; it is a vitamin K–dependent protein of bone. Although most of the synthesized osteocalcin accumulates in bone, a small fraction is secreted into the circulation. Catabolism of osteocalcin results in urinary excretion of γ-carboxy-glutamic acid (Gla), an amino acid. Both serum osteocalcin and urinary Gla are being used in the assessment of bone disease. Variables affecting blood levels of osteocalcin include (57):

- Osteoblastic synthesis.
- Content of Gla in the protein.
- Drug-induced alterations of osteocalcin affinity for bone.
- Hormonal status.
- Renal function.
- Age and sex; timing of specimen.
- Specificity of radioimmunoassay.

A strong correlation between serum bone Gla protein and serum alkaline phosphatase has been identified. Both indices are regarded as useful markers of bone formation (58).

Just as osteocalcin and alkaline phosphatase are biochemical markers of osteoblast activity (bone formation), so are fasting urinary calcium and the fasting urinary hydroxyproline:creatinine ratio clinical markers of osteoclast activity (bone breakdown). A fasting-state spot urine test that finds a urinary calcium:creatinine ratio exceeding 0.16 suggests high bone turnover (51). Calciuria in excess of 150 mg/day is indicative of high-turnover osteoporosis (59).

These metabolic bone turnover indices are useful in differentiating the cause of osteoporosis. In early postmenopausal osteoporosis, both bone resorption and compensatory bone formation are increased; in alcohol-induced osteoporosis, bone resorption is normal or increased while bone formation is impaired (60). Monitoring of these biochemical indices has proved useful not only for diagnosing the early presence and the type of osteopenia, but also for monitoring the therapeutic response and predicting intervention outcomes.

In postmenopausal women on estrogen therapy, an increase in bone mineral content may only occur in those who demonstrate a gradual decline in serum alkaline phosphatase levels accompanied by a rapid decline in hydroxyproline. The difference between hydroxyproline (a bone resorption index) and alkaline phosphatase (a bone formation index) reflects the calcium balance and hence bone mineral content. In estrogen-supplemented postmenopausal women, the decline in hydroxyproline levels is believed to reflect the skeletal response to estrogen, and it is therefore postulated that estrogen supplementation should be initiated early in order to impede bone resorption (61). Analysis of biochemical indices suggests that estrogen impairs bone resorption rather than enhancing bone formation; early estrogen replacement is therefore essential. In postmenopausal osteoporosis, the drop in estrogen level that heralds the onset of exponential bone loss is accompanied by slight but significant increases in plasma and urinary calcium, urinary hydroxyproline, and alkaline phosphatase. The increase in these biochemical indices in the posttreatment state does not significantly exceed pretreatment or placebo values. Based on these data, the researchers concluded that withdrawing estrogen/gestagen therapy at this stage in postmenopausal women is unlikely to initiate a rebound acceleration of bone loss (62).

Monitoring biochemical indices in conjunction with dual photon absorptiometry and/or total neutron activation analysis can provide precise information about bone remodeling activity and current bone status (63,64). Anatomical data provide information about the structural stage to which the disease has progressed; biochemical information supplies insight into the dynamic pathophysiological changes taking place. The more precisely pathophysiological changes can be defined in a given patient, the more accurate the diagnosis and the more appropriate the selected intervention strategy.

CLINICAL INTERVENTION: A PATHOPHYSIOLOGICAL APPROACH

Three determinants have been found to influence the pathogenesis of osteoporotic fractures:

- Bone mass. When bone density falls below 1.0 g/cm^2, the fracture threshold is breached and fractures occur as a result of minimal trauma (65).
- Bone strength. Unrepaired fatigue; microdamage; altered architecture and bone distribution, including loss of the connectivity and porosity of a thin cortex; impaired material properties, such as increased brittleness, molecular disorganization, and defective mineralization—all may impair bone strength and predispose to fractures.
- The propensity to fall. Regular exercise and considered modification of the person's structural environment can markedly reduce the risk of falling.

While all three areas are possible targets for osteoporosis prevention, the area most frequently targeted in skeletal health promotion is maximization and maintenance of an adequate bone mass. The aim of this health promotion strategy is to achieve a large peak bone mass before the third decade and avoid accelerated postmenopausal bone loss. Nutrition, exercise, and hormonal replacement are all possible components of a bone mass maximization and maintenance program (66).

Bone reaches its maximum mineral density about ten years after the cessation of linear growth. Peak bone mass is usually achieved before the age of 35 years; after this time, age-related bone loss is progressive (67). The greater the peak bone mass achieved by an individual, the larger the "buffer zone" between actual or current bone mass and the fracture threshold. Peak bone mass is probably the single most important determinant of an individual's risk of a fracture in later life. Bone mineral density begins to decline during the third decade and reaches about half of its maximum value at about 80 years. Cortical bone loss probably begins after 40 years of age and then accelerates for five to ten years after menopause. This accelerated bone loss continues for up to 15 years before gradually slowing to the premenopausal cortical bone loss rate. It has been estimated that women lose 35% of cortical and 50% of trabecular bone as they age; a man loses two-thirds of this amount. Loss in trabecular bone with aging is associated with complete perforation and fragmentation of some trabeculae. Most investigators suggest that the decline in trabecular bone mass precedes cortical bone loss starting after the age of 30; others believe that trabecular bone mass only declines after cessation of ovarian function. In any event, after the age of 30 years the overall balance between bone resorption and bone formation is altered to favor bone resorption. Reduced bone mass resulting from such an imbalance may be attributable to either absolute or relative increases or decreases in bone formation and resorption. Bone mass is determined by the relative balance between these two processes. Calcium intake and exercise habits are two major variables believed to influence bone mass.

There is some controversy about the scientific verification of both exercise and calcium supplementation as efficacious intervention measures. Further investigation is required, particularly on the effect of exercise and calcium supplementation on achieving and maintaining peak bone mass. Current knowledge does not adequately support the prescription of exercise and calcium supplementation as means of preventing the progression of established osteoporosis; neither do clinical trials conclusively refute such a hypothesis. Clinical trials show conflicting results (41).

Exercise

Bone thickness and strength are dependent on mechanical forces applied to the skeleton. Weight bearing exercise can promote and preserve bone strength. Bone mass is maximized and maintained when bone is subject to stress, and it has been proposed that bone mass may actually increase when bone is physiologically stressed in the presence of adequate skeletal nutrition (68). Exercise can certainly modify bone mass. Inadequate calcium intake in turkeys was found to result in loss of bone mass; this loss was exaggerated in disuse and reduced with exercise (39). In humans, astronauts experience bone loss when subjected to weightlessness, and moderate exercise has been shown to modify involutional bone loss (69); total body calcium was increased in the exercise group even when no calcium supplementation was prescribed beyond that present in the subject's normal diet. A different research group confirmed these findings in a controlled trial showing that exercise could inhibit involutional lumbar vertebral bone loss (70).

A nonrandomized controlled clinical trial examining lumbar bone mineral content in women supplemented with 1.5 g of calcium daily found that weight bearing exercise significantly increased bone mass after nine months (5.2% above baseline), with no change in the control group (−1.4%); this increased slightly over the next 13 months (6.1% above baseline). After 13 months of decreased activity, lumbar bone mass had reverted to 1.1% above the baseline level. The authors concluded that weight bearing exercise leads to significant increases of bone mineral content in postmenopausal women when continuous training is maintained (71). The exercise used in this study consisted of weight bearing, e.g., jogging, walking, or stair climbing at 70–90% of maximal oxygen intake for 50–60 minutes three times a week. A controlled clinical trial of healthy postmenopausal women suggests that aerobic exercise with or without strengthening exercises does modify bone loss (72). Clinical trials have shown the following results (73):

- There is an increase in total body calcium in exercising women compared with nonexercising women.
- Runners have about 40% more bone mass than matched controls.
- Lumbar bone mass can increase in women between 50–73 years of age with weight bearing exercise.
- Compared with nonexercising controls, exercise of 30–60 minutes on three days of the week resulted in statistically significant increases in bone mass within two to three years of initiating this intervention, as measured by single photon absorptiometry of the forearm.
- In one small trial, a statistically significant increase in total body calcium was found following 60 minutes of exercise on three days of the week over a 12 month period.

Although these studies do support a role for exercise, the conclusion of a substantial literature review is that few empirical data exist to support assertions that exercise can have a significant and lasting effect on maximizing and maintaining bone mass or retarding postmenopausal and senile osteoporosis (74). The reviewers concluded that the results of observational studies suggest that the extent of exercise performed and the site at which bone mass is determined alter clinical findings. A difficulty inherent in such cross-sectional studies is the omission of preexercise bone mass determinations. Until an adequate number of studies using randomized designs have been performed and published, methodological flaws will impede a conclusive decision about the usefulness of exercise in osteoporosis prevention (75).

It is currently not possible to make confident recommendations for the prophylactic use of exercise; neither the type nor the amount of exercise that will optimally maximize and maintain peak bone mass has been satisfactorily identified. Nonetheless, in the absence of more precise data, weight bearing exercise for 30–45 minutes three or four times a week seems a reasonable recommendation for skeletal health (76). This exercise should preferably take place outdoors so that vitamin D synthesis can be enhanced. As the change in bone mineral content is an adaptation to the mechanical stress of exercise on bone, persistent exercise of a particular nature will most likely lead to a steady state in which activity balances bone mass. The total remodeling time for a bone multicellular unit is of the order of four to six months; a training of up to 12 months may therefore be required before a steady state is reached. Further increases in bone mass are then likely to require higher levels of bone stress or exercise. Bone adaptation to the stresses imposed by exercise are probably limited to those bones stressed by the particular activity. Research should therefore be performed to identify which exercises will achieve increased bone mass in bones at risk; it is also important to identify the extent and duration of the required exercise. In view of the proven bone loss associated with disuse atrophy, any bone mass increases produced by exercise will probably be lost once exercise is discontinued. In the absence of adequate calcium levels (over 750 mg/day), exercise may be ineffective in modifying bone mass (76).

Calcium

The resistance of bone to fracture is a function of its mineral content and its trabecular structure; the strain that a bone can sustain before fracturing can be indirectly estimated by measuring the bone's mineral content. Analysis of the mineral content of the skeleton suggests that calcium is a predominant nutrient. Calcium is an important mineral in the hydroxyapatite lattice of bone; chronic negative calcium balance can manifest as a reduction of bone mass. In such a situation, calcium is leeched from bone in order to maintain normal serum calcium levels. Although not yet conclusively proven, calcium intake in early life probably influences peak bone mass. It may be considered prudent to at least ensure that calcium intake meets the RDA level. This is not always the case (77,78).

Skeletal health promotion may be attempted by restoring calcium intake to RDA levels, but prevention of osteoporosis by boosting calcium intakes to 50% above RDA levels is more controversial. In this context, the 1984 National Institutes of Health consensus report explicitly recommended calcium supplementation, whereas the 1986 research findings of an eminent research hospital concluded that calcium is the laetrile of osteoporosis (79). Gordon and Vaughan (38) consider the proposition that megadoses of calcium premenopausally may prevent bone loss postmenopausally an "amazing concept." They explain their skepticism about the value of calcium supplementation as a means of maintaining bone mass by quoting, amongst others, the work of Mazess and coworkers (80) and Gennari and coworkers (81). Calcium supplementation that goes beyond implementing accepted dietary guidelines and/or proposes calcium intakes beyond that recommended by the health authorities requires increased justification—on both clinical and scientific grounds. At its 104th session, the National Health and Medical Research Council (82) of Australia noted that "Recent studies had not confirmed any benefit of taking calcium supplements to prevent osteoporosis"; it also remarked that many sources were advising the community on the role of calcium in the prevention of osteoporosis and that "much of this advice is confusing and often misleading." This recommendation disputes the conclusion of the NIH Consensus Panel (83) that strategies in the prevention of osteoporotic fractures include "ensuring estrogen replacement in postmenopausal women, adequate nutrition including elemental calcium intake of 1,000 to 1,500 mg/day, and a program of modest weight bearing exercise." A prospective sequential clinical trial found that calcium supplementation appeared to benefit the appendicular skeleton in postmenopausal women (84). A similar study assessing the effect of calcium supplementation on the axial skeleton would be most helpful. Despite probably minor benefit to cortical bone, no benefit on trabecular bone was documented to arise from 2,000 mg/day of calcium carbonate (85).

One explanation for the apparent lack of response of certain postmenopausal women to calcium supplementation may be related to failure to absorb dietary calcium. In osteoporotic women with normal calcium absorption, calcium was able to suppress bone resorption; if calcium was abnormal, a combination of calcium and calcitriol was required to obtain a similar result (86). At least one double-blind, controlled clinical study contradicts this result by finding that daily calcium supplementation of 2.0 g in postmenopausal women did not prevent trabecular bone loss (85); calcium combined with vitamin D did not improve this result (87). In contrast, biopsy samples of radial and phalangeal bone in persons with senile osteoporosis revealed an improved mineralization response both to calcium alone and to a combination of calcium and calcitriol (88). Other studies in postmenopausal women found that high doses of calcium were only effective when taken in combination with estrogen (89).

An additional dimension to the calcium controversy is encountered when the issue of calcium supplementation in osteoporosis prevention is expanded to include consideration of the dose and duration of therapy. Researchers have suggested that maintenance of high-dose calcium supplements (2,500 mg/day) improves calcium balance for three to 36 months; more recently it has been found that these effects diminish significantly after six months (90). In his recent review, Bauwens (91) suggests that in spite of inadequate follow-up (less than three years in most cases) and an emphasis on cortical bone, six studies published between 1977 and 1982 do seem to suggest some clinical support for calcium at least partially reducing the rate of bone loss when at least 1,500 mg/day are ingested by women who have already lost substantial quantities of bone or have sustained a fracture. He also points out, however, that two recent studies fail to find any correlation between calcium intake and slowed bone loss as measured by densitometry.

Because dietary calcium is more likely to suppress PTH and increase calcitonin secretion than to have any direct effect on osteoblast or osteoclast function, it is more likely to be of assistance in achieving a substantial peak bone mass than protecting against rapid bone loss.

A calcium intake of 1,000 mg/day is recommended for men and premenopausal women; this should be increased by 1,500 mg/day in postmenopausal women. If a calcium supplementation is contemplated, this should be ingested at night, and one should be aware of the concentration of elemental calcium present. Calcium carbonate contains 40% elemental calcium; calcium gluconate, 9% elemental calcium; calcium lactate, 13% elemental calcium; and dibasic calcium phosphate, 23% elemental calcium. Dairy products are the best food sources of calcium; calcium in vegetables and other fiber-rich foods is often bound to phytates and oxalates.

Other dietary measures that should be considered are those targeting calcium excretion. Calcium-sparing intervention measures include avoiding excessive consumption of high-protein foods, salt, alcohol, and caffeine-based drinks. Nicotine usage should also be decreased.

Managing Established Osteoporosis

In attempting to maximize the individual's potential to achieve a large peak bone mass, it is logical, and possibly rational, to provide growing bones with a mechanical and nutritional environment conducive to skeletal health. When attempting to prevent bone loss, a more careful scrutiny of bone remodeling is proposed. Therapeutic prevention of osteoporosis requires a cognizance of both bone mass and bone strength.

Some individuals with osteoporosis have increased osteoclastic resorption with increased rates of bone turnover—a high bone turnover state; others have decreased bone formation associated with decreased remodeling—a low bone turnover state. Increased osteoclastic bone resorption dominates the early postmenopausal period; decreased osteoblastic function is the primary defect of the older patient. Low-turnover states are particularly prevalent in the elderly, in alcoholics, and in persons with Cushing's syndrome. Low-turnover states should be treated with drugs that stimulate bone formation, high-turnover states with drugs that inhibit bone resorption. Sodium fluoride and PTH stimulate bone formation; calcium, estrogen, calcitonin, and calcitriol inhibit bone resorption. Drugs that act by suppressing bone resorption probably produce a modest increment in bone mass by filling the remodeling space. An increase in bone mass or density can be achieved by drug intervention in established osteoporosis; better elucidation of the relationship between bone mass and bone strength is, however, desirable. Although presently a theoretical concern, the question as to whether long-term suppression of bone remodeling inhibits microfracture repair to the detriment of bone quality is unanswered (92).

Currently, there are two theoretical approaches to treatment based on the fracture threshold hypothesis. Drugs that inhibit bone resorption will prevent further bone loss by limiting bone resorption—estrogen therapy is based on this perspective. In contrast, drugs stimulating bone formation should increase bone density above the fracture threshold—this is the basis of fluoride therapy. Interventions designed to inhibit bone resorption may be limited to a short-term effect. Recent evidence suggests that drugs that limit resorption will not increase bone density above the fracture threshold and, once a new bone-remodeling steady state has been established, bone formation may decline to a level at which bone loss resumes at a rate similar to that preceding therapy. Drugs that stimulate bone formation have been shown to increase bone mass; any associated increase in bone strength has yet to be demonstrated. One recent attempt to address the mushrooming dilemma of osteoporosis offers an integrated therapy in the form of an ADFR (Activation-Depression-Free-Repeat) regimen (43,93). This is founded on the hope that bone remodeling may be better stimulated by using, in sequence, drugs that activate new bone multicellular units; drugs that inhibit activation of newly generated osteoblasts but not osteoclasts; and a drug-free phase, during which the new bone multicellular units, now overfilled with osteoblasts, will produce a physiologically strong bone matrix. The sequence is then repeated. Further clinical validation is required.

Hormone Replacement Therapy

There is strong evidence to suggest that estrogen protects against increased bone loss in the immediate postmenopausal period; there is little evidence justifying its use in older patients with established disease. The rationale underlying estrogen therapy is based on the understanding that estrogen can slow the rapid bone loss (cortical and trabecular) that occurs for about ten years after menopause; retrospective studies of patients with hip and vertebral fractures show that these fractures occur less commonly in patients who have taken estrogen for at least five years following menopause. There is therefore a substantial argument for use of estrogen as a prophylactic measure in perimenopausal women. Criteria for the prescription of prophylactic hormone replacement include:

- The presence of more than four risk factors in a menopausal woman; risk factors include a small-boned slender body build; a sedentary lifestyle; a low-calcium, high-salt, and high-protein diet; smoking; and disorders including diabetes mellitus, rheumatoid arthritis, and malabsorption syndromes.
- The ability to initiate therapy within 12–24 months of menopause.
- An absence of contraindications to estrogen therapy; contraindications include hormone-dependent breast cancer, a history of uterine cancer, estrogen-related hypertension, triglyceridemia, or a history of thromboembolic disease.
- A low bone mass as measured by photon absorptiometry or equivalent measurement. This criterion is only required by certain clinicians.

Estrogen is prescribed for ten to 12 years. Premarin 0.625 mg/day for 21 days is supplemented with progestin for ten days in women with an intact uterus. Six-monthly evaluation for occult uterine cancer is required. Side effects include an increased risk of endometrial and/or breast cancer, gallbladder disease, hypertension, and hypertriglyceridemia. Minor side effects include nausea, vomiting, bloating/edema, headaches, breast tenderness, recurrent vaginal bleeding, and increased frequency of abnormal vaginal bleeding. Serendipitous advantages include fewer hot flushes, reversal of vaginal atrophy, and, possibly, a reduced risk of ischemic heart disease.

Estrogen replacement is often regarded as the definitive treatment for female osteoporosis (94). Physiological estrogen withdrawal certainly results in increased osteoclastic bone resorption. Hormone replacement is probably indicated in peri- and postmenopausal women with osteopenia (95). Many experts believe that estrogen replacement is the most effective therapy for preventing the rapid loss of bone mass following the onset of menopause; estrogen should therefore be administered in the perimenopausal period as a prophylactic measure rather than delayed for use as a therapy in established disease.

Fluoride

Skeletal fluorosis, a condition of hyperossification, has long been recognized when excess fluoride is ingested in the absence of adequate calcium. The first manifestation is radiological evidence of increased bone density; osteosclerosis is most marked in the vertebrae and pelvis. Skeletal fluorosis has been described as passing through the phases characterized by patchy osteoporosis, patchy osteosclerosis/hypermineralization, osteomalacia with bands of uncalcified osteoid tissue, and finally severe chronic osteofluorosis. On microscopy this manifests as a mixture of normal bone and dense (sclerotic), rarefied (porotic), and disorganized (osteomalacia) bone. Exostoses and osteophytes are radiologically detectable. Calcification of ligaments and tendons can manifest clinically as a poker spine or hunch back. Radiological evidence of fluorosis is found in the form of osteosclerosis in 10–15% of people who consume 8 ppm fluoride for prolonged periods. It is the predilection for 95% of ingested fluoride to deposit in bone that has raised the possibility of fluoride having a role in the management of osteoporosis.

"The ideal therapeutic program for osteoporosis is based not on the slowing of bone loss but rather on making positive the bone tissue balance"; this is not achieved by calcium and vitamin D alone, but it can result when fluoride is added (96,97). Apart from hormonal therapy, the use of sodium fluoride, in conjunction with recommended calcium and physiological vitamin D levels, appears to be the only realistic alternative capable of increasing trabecular thickness in osteoporotic bones (98). While fluoride is considered to "probably be an essential nutrient," no RDA level has been determined. The clinical use of this mineral should therefore be analyzed within the management and research models applied to drug therapy.

Sodium fluoride, unlike calcium, is a potent stimulator of bone formation; fluoride can actually increase bone mass in osteoporotic bones (98). Fluoride enhances bone formation by direct stimulation of osteoblastic activity. It is therefore particularly useful in patients with decreased bone turnover and impaired bone formation. Fluoride stimulates osteoid formation and is itself incorporated into the hydroxyapatite of bone, thereby altering its crystalline structure and increasing bone density, but not necessarily bone strength. The fluoride ion substitutes for the hydroxyl radicals in the hydroxyapatite crystals to form fluorapatite, which contributes to increased bone crystallinity and decreased solubility. Fluoride produces a positive uncoupling that favors formation over resorption with a consequent increase in bone mass. Trabecular bone, especially of the axial skeleton, is responsive to fluoride; cortical bone may be unaffected. While fluoride administration in conjunction with calcium and vitamin D appears to be associated with lowered fracture rates, it must be emphasized that if administered alone fluoride may result in abnormal bone formation and increased fractures (99,100). Adequate calcium is essential for mineralization of the osteoid produced in response to fluoride stimulation. Appropriate use of fluoride in osteoporosis management presupposes associated calcium supplementation (98); fluoride administered in the presence of calcium deficiency results in unmineralized bone formation. To improve absorption, fluoride and calcium should be administered separately in daily doses of 40–50 mg of sodium fluoride and 1,000 mg of calcium. A daily dose of 30 mg of sodium fluoride combined with 1 g of calcium has been found to increase spinal bone mineral content in a prospective controlled study (101). Another prospective study ascertaining the risk-benefit ratio of sodium fluoride suggests that a reasonable compromise between antifracture effectiveness and side effects is achieved by daily doses of 50 mg of sodium fluoride (96).

At therapeutic dose levels, side effects attributable to fluoride include nausea, vomiting, diarrhea, abdominal pain, and occasionally gastrointestinal bleeding. Rheumatic symptoms include plantar fascial pain, periarticular pin, synovitis, and arthritis. Symptoms disappear some three weeks after withdrawing fluoride. About one in three patients consider fluoride therapy intolerable. Analysis of a five-year program for therapy of osteoporotic vertebral crush fractures with sodium fluoride, calcium phosphate, and vitamin D showed that over half of the patients experienced joint-related or gastrointestinal side effects that subsided after a median six-week withdrawal of fluoride (102). Absolute contraindications to fluoride therapy include renal failure and osteomalacia (fluoride forms spongy and irregular bone in the absence of adequate calcium and vitamin D) (93).

The therapeutic index of fluoride is about 100 to 1. This is the ratio between the dosage effective in reversing osteoporotic lesions in the elderly and that causing toxicity in persons who are not forming teeth. This level of safety is about the same as for aspirin in relieving a headache (103). Despite fluoride intervention being considered "experimental," reputable physicians include daily administration of 40–60 mg of sodium fluoride as an option in the management of osteoporosis (104,105). The NIH Consensus Panel's (83) comment is that while sodium fluoride, in association with high calcium intake, may have a role to play in the management of patients with severe osteoporosis, its efficacy and safety remain unproven.

ASSESSING THE EFFICACY OF OSTEOPOROSIS MANAGEMENT

Despite extensive research, no definitive recommendations have been issued on optimal regimens for either increasing peak bone mass or managing established osteoporosis. Before adequate recommendations can be issued, sufficient publications must take cognizance of a number of natural history variables. Variables in the pathogenesis of osteoporosis that have been identified as affecting the reliability and validity of intervention strategies include:

- The site of bone density measurements. Bone density should be assessed at all key sites. The kinetics of cortical and trabecular bone are different.
- The duration of patient follow-up in prospective studies. At least three years of follow-up are required to produce data that may have predictive value.

- The heterogeneity of the condition loosely labeled as involutional osteoporosis. Histological differentiation may be helpful in distinguishing between senile and postmenopausal osteoporosis and osteoporosis associated with secondary hyperparathyroidism and impaired bone formation.
- Adequate definition of the study population. Standardized inclusion criteria should be established for all clients; e.g., inclusion criteria used for spinal deformity range from one 15% anterior wedge to two 25% anterior wedges, suggesting different client populations.

Current frequently used indices include:

- Fracture rate. "New" fractures are defined by radiological changes based on vertebral body shape. There is, however, no evidence that these early changes are related to later fractures.
- Serial measures of bone mineral density/bone mass. The patient should be observed for at least three years. Any agent inhibiting bone resorption should produce detectable differences between subjects and controls within three to 12 months. However, following this a steady state is obtained, rates of bone formation will slow to less than rates in those with diminished bone resorption, and bone loss will resume at rates existing prior to therapy. Only prolonged observation can lead to valid conclusions.
- Pain. At best pain is an unreliable indicator of bone status; at worst it is a reflection of ethnic and cultural rather than physical status.

Only adequate standardization of research protocols can facilitate a meaningful comparison of research findings and consequent promulgation of clinically valid intervention strategies.

When postulated intervention is scientifically inconclusive, any proposed alteration to a population's behavior should seek to normalize rather than innovate. In such circumstances it is preferable to restore biological normality by removal of abnormal exposures, e.g., smoking, and moderation of recently acquired dietary habits than to impose some new, supposedly protective intervention, e.g., drugs or jogging (4). Although such benign intervention may constitute little more than placebo therapy, it is a relatively safe option when the pathogenesis of a condition is largely unproven and intervention uncertain. The utility of causal diagnosis in management decision making is directly proportional to the precise understanding of the natural history of the presenting condition.

References

1. McDonnell JM, Lane JM, Zimmerman AB: Osteoporosis: definition, risk factors, etiology, and diagnosis. *AAOHN J* 35:527–530, 1987.
2. Genant HK, Cann CE, Ettinger B, et al: Quantitative computed tomography of vertebral spongiosa: a sensitive method for detecting early bone loss after oophorectomy. *Ann Intern Med* 97:699–705, 1982.
3. Kelsey J: *Research Direction in Osteoporosis*. Washington, DC, National Institutes of Health, 1987.
4. Rose G: Sick individuals and sick populations. *Int J Epidemiol* 14:32–38, 1985.
5. Rudy DR: Osteoporosis. *Postgrad Med* 86:151–158, 1989.
6. McDermott MT, Kidd GS: The role of calcitonin in the development and treatment of osteoporosis. *Endocr Rev* 8:377–390, 1987.

7. Briancon D, Meunier PJ: Non-hormonal treatment of postmenopausal osteoporosis. *Rev Fr Gynecol Obstet* 82:713–718, 1987.
8. MacIntyre I, Stevenson JC, Whitehead MI, Wimalawansa SJ, Banks LM, Healy MJR: Calcitonin for the prevention of postmenopausal bone loss. *Lancet* 1:900–901, 1988.
9. Jaworski ZFG: Physiology and pathology of bone remodeling, *Orthop Clin North Am* 12:485–512, 1981.
10. Nordin BEC, Need AG, Morris HA, Horowitz M: New approaches to the problems of osteoporosis. *Clin Orthop Rel Res* 200:181–197, 1985.
11. Nordin BEC: The calcium debate. *Med J Aust* 148:608–609, 1988.
12. Nordin BEC, Polley KJ, Need AG, Morris HA, Marshall D: The problem of calcium requirement. *Am J Clin Nutr* 45:1295–1304, 1987.
13. National Health and Medical Research Council. *Dietary Allowances for Use in Australia.* Canberra, Australian Government Publishing Service, 1979.
14. Editorial. Risk factors in postmenopausal osteoporosis. *Lancet* 1:1369–1372, 1985.
15. Heaney RP, Recker RR, Saville PD: Menopausal changes in calcium balance performance. *J Lab Clin Med* 92:953–963, 1978.
16. Heaney RP, Recker RR: Distribution of calcium absorption in middle aged women. *Am J Clin Nutr* 43:299–305, 1986.
17. Parfitt AM: Dietary risk factors for age related bone loss and fractures. *Lancet* 2:1181–1185, 1983.
18. Angus RM, Pocock NA, Eisman JA: Nutritional intake of pre- and postmenopausal Australian women with special references to calcium. *Eur J Clin Nutr* 42:617–625, 1988.
19. Angus RM, Eisman JA: Osteoporosis: the role of calcium intake and supplementation. *Med J Aust* 148:630–633, 1988.
20. Briggs D, Wahlqvist M: *Food Facts*. Ringwood, Penguin Books Australia, 1985.
21. Recker RR, Heaney RP: The effect of milk supplements on calcium metabolism, bone metabolism and calcium balance. *Am J Clin Nutr* 41:254–263, 1985.
22. *Report of the Royal College of Physicians. Medical Aspects of Dietary Fibre.* Tunbridge Wells, Pitman Medical, 1980.
23. Wardlaw G: The effects of diet and life-style on bone mass in women. *J Am Diet Assoc* 88:17–25, 1988.
24. Sabto J, Powell MJ, Breidahl MJ, Gurr FW: Influence of urinary sodium on calcium excretion in normal individuals. *Med J Aust* 140:354–356, 1984.
25. Goulding A, Lim PE: Effects of varying dietary salt intake on the fasting urinary excretion of sodium, calcium and hydroxyproline in young women. *N Z Med J* 96:853–854, 1983.
26. Goulding A, Everitt HE, Cooney JM, Spears GFS: Sodium and osteoporosis. In Wahlqvist ML, Truswell AS (eds): *Recent Advances in Clinical Nutrition*. London, John Libby, 1986, vol II, pp 99–108.
27. Beauchamp GK, Bertino M, Engelman K: Modification of salt taste. *Ann Intern Med* 98:763–769, 1983.
28. Breslau NA, McGuire JL, Zerwekh JE, Pak CYC: The role of dietary sodium on renal excretion and intestinal absorption of calcium and on vitamin D metabolism. *J Clin Endocrinol Metab* 55:369–373, 1982.
29. Nordin BEC: Calcium. *J Food Nutr* 42:67–82, 1986.
30. Marcus R: The relationship of dietary calcium to the maintenance of skeletal integrity in man—an interface of endocrinology and nutrition. *Metabolism* 31:93–102, 1982.
31. Yuen DE, Draper HH, Trilok G: Effect of dietary protein on calcium metabolism in man. *Nutr Abstracts Rev Ser A* 54:447–459, 1984.
32. Linkswiler HM, Joyce CL, Anand CL: Calcium retention of adult males as affected by level of protein and of calcium intake. *Trans N Y Acad Sci Ser II* 36:333–340, 1974.
33. Lutz J: Calcium balance and acid-base status of women affected by increased protein intake and by sodium bicarbonate ingestion. *Am J Clin Nutr* 39:281–288, 1984.
34. McDermott MT, Wittee MC: Bone mineral content in smokers. *South Med J* 88:17–25, 1988.
35. Finn S: Osteoporosis and nutrition. *AAOHN J* 35:536–538, 1987.
36. Albanese AA, Lorenze J, Wein EH: Osteoporosis: effects of calcium. *Am Fam Physician* 18:160–167, 1978.

37. Lanyon LE, Rubin CT, Baust G: Modulation of bone loss during calcium insufficiency by controlled dynamic loading. *Calcif Tissue Int* 38:209–216, 1986.
38. Gordon GS, Vaughan C: Calcium and osteoporosis. *J Nutr* 116:319–322, 1986.
39. Woolf AD, Dixon AStJ: Osteoporosis—an update on management. *Drugs* 28:565–576, 1984.
40. Cummings SR, Kelsey JL, Nevitt MC, Oldowd KJ: Epidemiology of osteoporosis and osteoporotic fractures. *Epidemiol Rev* 7:178–208, 1985.
41. Avioli LV: Postmenopausal osteoporosis: prevention versus cure. *Fed Proc* 40:2418–2422, 1981.
42. Dawson-Hughes B, Jacques P, Shipp C: Dietary calcium intake and bone loss from the spine in healthy postmenopausal women. *Am J Clin Nutr* 46:685–687, 1987.
43. Mundy GR: Osteopenia. *Dis Mon* 33:537–600, 1987.
44. Ettinger B: Estrogen and postmenopausal osteoporosis. *AAOHN J* 35:543–546, 1987.
45. Bilbrey GL, Weix J, Kaplan GD: Value of single photon absorptiometry in osteoporosis screening. *Clin Nucl Med* 13:7–12, 1988.
46. Burton VJ, Rob MI, Craig GC, Lawson JS: Changes in caries experience of 12-year-old Sydney schoolchildren between 1963 and 1982. *Med J Aust* 140:405–407, 1984.
47. Ott SM, Kilcoyne RF, Chesnut CH: Comparison among methods of measuring bone mass and relationship to severity of vertebral fractures in osteoporosis. *J Clin Endocrinol* 66:501–507, 1988.
48. Zilkoski M, Morrow LB: Osteoporosis. *Am Fam Physician* 36:178–186, 1987.
49. Rippe JM, Ward A, Porcari JP, Freedson PS: Walking for health and fitness. *JAMA* 259:2720–2724, 1988.
50. Nilas L, Gotfredsen A, Riis BJ, Christiansen C: The diagnostic validity of local and total bone mineral measurements in postmenopausal osteoporosis and osteoarthritis. *Clin Endocrinol (Oxf)* 25:711–720, 1986.
51. Bellantoni MF, Blackman MR: Osteoporosis: diagnostic screening and its place in current clinical care. *Geriatrics* 43:63–66, 69–70, 1988.
52. Genant HK, Cann CE: Clinical impact of quantitative computed tomography for vertebral mineral assessment. In Marqulis AR, Goodman CA (eds): *Diagnostic Radiology: 26th Postgraduate Course*. San Francisco, University of California Press, 1983, pp 445–448.
53. Vogler JB, Martinez S: Non-invasive procedures: pros and cons. *AAOHN J* 35:547–549, 1987.
54. Genant HK, Block JE, Steiger P, Glueer CC, Smith R: Quantitative computed tomography in assessment of osteoporosis. *Semin Nucl Med* 17:316–333, 1987.
55. Doren M, Montag M, Schneider HP: Postmenopausal osteoporosis. Early diagnosis as an indication for preventive treatment. *Radiologe* 28:149–152, 1988.
56. Nordin BEC, Need AG: Diagnosis and management of osteoporosis. *Aust Fam Physician* 17:170–173, 1988.
57. Lian JB, Gundberg CM: Osteocalcin. Biochemical considerations and clinical applications. *Clin Orthop* 226:267–291, 1988.
58. Podenphant J, Christiansen C, Catherwood BD, Deftos LJ: Serum bone Gla protein and other biochemical estimates of bone turnover in early postmenopausal women during prophylactic treatment for osteoporosis. *Acta Med Scand* 218:329–333, 1985.
59. Nordin BEC, Crilly G, Smith DA: Osteoporosis. In Nordin BEC (ed): *Metabolic Bone and Stone Disease*, 2nd ed. Edinborough, Churchill Livingstone, 1984, pp 1–70.
60. Johansen JS, Riis BJ, Delmas PD, Christiansen C: Plasma BPG: an indicator of spontaneous bone loss and of the effect of oestrogen treatment in postmenopausal women. *Eur J Clin Invest* 18:191–195, 1988.
61. Christiansen C, Rodbro P, Tjellesen L: Serum alkaline phosphatase during hormone treatment during early postmenopausal women. A model for establishing optimal prophylaxis and treatment in postmenopausal osteoporosis. *Acta Med Scand* 216:11–17, 1984.
62. Thomsen K, Riis BJ, Johansen JS, Christiansen C, Rodbro P: Bone turnover in postmenopausal women after withdrawal of oestrogen/gestagen replacement therapy. *Gynecol Endocrinol* 1:169–175, 1987.
63. Cohn SH, Aloia JF, Vaswani AN, Yuen K, Yasumura S, Ellis KJ: Women at risk for

developing osteoporosis: determination by total body neutron activation analysis and photon absorptiometry. *Calcif Tissue Int* 38:9–15, 1986.
64. Prince R: How should osteoporosis be diagnosed? *Patient Manage* 12:31–40, 1988.
65. Riggs BL, Wahner HW, Melton JL, et al: Differential changes in the bone mineral density of the appendicular and axial skeletons of women: evidence of substantial vertebral bone loss prior to menopause. *J Clin Invest* 67:328–335, 1981.
66. Kirkpatrick MK: A self-care model for osteoporosis. *AAOHN J* 5:531–535, 1987.
67. Dixon AStJ: Osteoporosis—an unheeded epidemic. *Practitioner* 230:363–371, 1986.
68. Korcok M: Add exercise to calcium in osteoporosis prevention. *JAMA* 247:1106–1112, 1982.
69. Aloia JF, Cohn SH, Ostuni JA, Cane R, Ellis K: Prevention of involutional bone loss by exercise. *Ann Intern Med* 89:356–358, 1987.
70. Krolner B, Toft B, Nielson SP, Tondevold E: Physical exercise as prophylaxis against involutional vertebral bone loss: a controlled trial. *Science* 64:541–546, 1983.
71. Dalsky GP, Stocke KS, Ehsani AA, Slatopolsky E, Lee WC, Birge SJ: Weight bearing exercise training and lumbar bone mineral content in postmenopausal women. *Ann Intern Med* 108:824–828, 1988.
72. Chow R, Harrison JE, Notarius C: Editorial. *Br Med J [Clin Res]* 295:1441–1444, 1987.
73. Goodman CE: Osteoporosis and physical activity. *AAOHN J* 35:539–542, 1987.
74. Block JE, Smith R, Black D, Genant HK: Does exercise prevent osteoporosis? *JAMA* 257:3115–3117, 1987.
75. Dalsky GP, Stocke KS, Ehsani AA, Slatopolsky E, Lee WC, Birge SJ: Weight-bearing exercise training and lumbar bone mineral content in postmenopausal women. *Ann Intern Med* 108:824–828, 1988.
76. Dalsky GP: Exercise: its effect on bone mineral content. *Clin Obstet Gynecol* 30:820–832, 1987.
77. Amschler DH: Calcium intake—a lifelong proposition. *J Sch Health* 55:360–363, 1985.
78. Jamison JR, Geraghty B, Keating G, Livingstone K: Osteoporosis screening and prevention in the chiropractic clinic. *J Manip Physiol Ther* 5:391, 1988.
79. Bloch K: Concepts and approaches to scientific inquiry. *Am J Clin Nutr* 45:1054–1059, 1987.
80. Mazess RB, Harper AE, DeLuca H: Calcium intake and bone. *Am J Clin Nutr* 42:568–571, 1985.
81. Gennari C, Chierichetta SM, Bigazzi S, Gonnelli S, Ferrara R, Zacchei F: Comparative effects on bone mineral content of calcium and calcium plus salmon calcitonin given in two different regimes for postmenopausal osteoporosis. *Curr Therap Res* 38:455–464, 1985.
82. National Health and Medical Research Council. *Nutrition Policy Statements.* Canberra, Government Publishing Service, 1988.
83. Consensus Panel, National Institutes of Health. Osteoporosis, *JAMA* 252:799–802, 1984.
84. Polley KJ, Nordin BE, Baghurst PA, Walker CJ, Chatterton BE: Effect of calcium supplementation on forearm bone mineral content in postmenopausal women: a prospective sequential controlled trial. *J Nutr* 117:1925–1935, 1987.
85. Riis B, Thomsen K, Christiansen D: Does calcium supplementation prevent postmenopausal bone loss? A double blind controlled clinical study. *N Engl J Med* 316:173–177, 1987.
86. Nordin BEC, Need AG, Morris HA, et al: New approaches to the problem of osteoporosis. *Clin Orthop Rel Res* 200:181–197, 1985.
87. Riggs BL, Seeman E, et al: Effect of the fluoride/calcium regime on vertebral fracture occurrence in post-menopausal osteoporosis. *N Engl J Med* 306:446–450, 1982.
88. Orwoll ES, McClung MR, Oviatt SK, Recker RR, Wiegel RM: Histomorphometric effects of calcium or calcium plus 25-hydroxyvitamin D_3 therapy in senile osteoporosis. *J Bone Miner Res* 4:81–88, 1989.
89. Ettinger B, Genant HK, Cann CE: Postmenopausal bone loss is prevented by treatment with low dosage estrogen with calcium. *Ann Intern Med* 106:40–45, 1987.
90. Heaney RP, Gallagher JC, Johnston CC, et al.: Calcium nutriture and bone health in the elderly. *Am J Clin Nutr* 36:986–1013, 1986.
91. Bauwens SF: Efficacy of various therapeutic regimens. *Health Values* 11:37–43, 1987.

92. Aloia JF: Osteoporosis. Where are we in 1988? *AAOH J* 35:523–526, 1987.
93. Bauwens SF: Therapeutic regimes: implications and efficacy for treating osteoporosis. *AAOHN J* 35:550–554, 1987.
94. Jackson TK, Ullrich IH: Understanding osteoporosis. *Postgrad Med* 75:118–125, 1984.
95. Mamelle N, Meunier PJ, Dusan R, Guillaume M, Martin JL, Gaucher A, et al: Risk-benefit ratio of sodium fluoride therapy in primary vertebral osteoporosis. *Lancet* 2:361–365, 1988.
96. Hansson T, Roos B: The effect of fluoride and calcium on spinal mineral bone content: a controlled, prospective (3 year) study. *Calcif Tissue Int* 40:315–317, 1987.
97. Biancon D, Meunier PJ: Treatment of osteoporosis with fluoride, calcium, and vitamin D. *Orthop Clin North Am* 12:629–648, 1987.
98. Lane JM, Healey JH, et al: Treatment of osteoporosis with fluoride and calcium: effects on vertebral fracture incidence and bone histomorphology. *Orthop Clin North Am* 15:729–745, 1984.
99. Clifton-Bligh P, Eisman J, Nordin BEC, Fraser RC, Redi I: Osteoporosis—panel discussion of advances in our understanding of this potentially preventable disease. *Curr Ther* 26:25–43, 1985.
100. Jowsey J: Osteoporosis. *Postgrad Med* 60:75–79, 1976.
101. Mallette LE: Osteoporosis. *Postgrad Med* 72:271–287, 1982.
102. Hasling C, Nielsen HE, Melsen F, Mosekilde L: Safety of osteoporosis treatment with sodium fluoride, calcium phosphate and vitamin D. *Miner Electrolyte Metab* 13:96–103, 1987.
103. Richmond VL: Thirty years of fluoridation: a review. *Am J Clin Nutr* 41:129–138, 1985.
104. Posen S: Osteoporosis: a clinical approach to management. *Patient Manage* March:61–72, 1988.
105. Ingham J: Postmenopausal osteoporosis. *Patient Manage* April:47–60, 1978.

5
Diagnostic Labeling: A Camouflage for Ignorance?

Diagnosis is a naming game. Given a diagnosis, the practitioner can, with confidence, not merely provide a rough sketch of the patient's clinical presentation but also speculate about the treatment the patient is receiving. Diagnostic decision making is directed toward selecting an appropriate label for the patient's presenting complaint. Regardless of whether condition categorization has emphasized a probabilistic, pattern recognition, or causal approach, label selection determines the subsequent clinical management strategy. Disease classification serves to catalogue clinical knowledge and experience; once classified, conditions are subject to standardized intervention and monitoring procedures. The importance of accurate labeling must not be underestimated; the necessity of clearly defined criteria for allocating diagnostic labels is absolute. Precision in clinical diagnosis becomes apparent when one appreciates that certain conditions are diagnosed on the basis of a single sign. The detection of a single clinical sign can lead to the patient's acquisition of a diagnostic label with profound prognostic and therapeutic implications. Essential hypertension is an example of a covert disease that, once detected by the case finding clinician, can have a profound impact on the patient's lifestyle and perceived health status.

The relatively simple technology of a sphygmomanometer has made it possible to detect essential hypertension in asymptomatic individuals. Epidemiological studies have shown that risk of essential hypertension, a prevalent condition in western society [1], increases with age [2,3]. Based on probabilistic reasoning and sign recognition, apparently healthy persons are labeled as "suffering" from essential hypertension. The confidence with which such labels are applied does, however, camouflage profound uncertainty about the cause of this condition. Ignorance about disease etiology complicates but does not preclude justifiable clinical intervention. In spite of the idiopathic nature of its etiology and the silent progression of its pathogenesis, the risk of complications arising from essential hypertension increases with the degree and duration of blood pressure elevation. Complications arising from hypertension include cardiac failure, cerebrovascular accidents, and renal impairment; hypertension is itself a major risk factor for ischemic heart disease. In an attempt to prevent such target organ damage, health authorities have recommended that "healthy populations" undergo blood pressure measurements at two-yearly intervals (4). Case finding is being encouraged in order to detect evidence of a disease process in apparently healthy people, a disease process of unknown etiology but proven complications.

THE DIAGNOSTIC DECISION

The diagnosis of hypertension depends on the clinical detection of elevated blood pressure on three or more occasions. It is imperative that blood pressure levels are accurately recorded, as detection of three raised blood pressure readings results in the disease labeling of an apparently healthy person. Blood pressure manometry requires that certain rituals be observed (5).

With the relaxed or rested patient in the seated position, a cuff is applied with the bladder over the brachial artery. The diameter of the cuff should exceed the patient's midarm width by about 20%; a suitable cuff for adults is 13–15 cm wide and 30–35 cm long. The cuff is inflated and the palpatory systolic pressure determined by monitoring the disappearance of the radial pulse. After complete deflation of the cuff and a pause of at least 30 seconds, the cuff may be reinflated and the blood pressure assessed by auscultation. After inflating the cuff to a pressure 30 mm Hg above the level at which the palpated systolic pulse disappeared, the brachial artery is auscultated in the antecubital fossa while deflating the cuff at 2 mm Hg/sec. Systolic pressure is recorded at the appearance of the first Korotkoff sound, and diastole is determined by the disappearance (phase V) of the Korotkoff sounds. This procedure should be repeated after an interval of at least three minutes. The mean of the lowest two readings should be recorded. When the blood pressure in the two arms is compared, a difference of 5–10 mm Hg is within normal limits; the higher reading should be recorded. A change of posture is most likely to alter the blood pressure readings of elderly persons, pregnant women, and persons with orthostatic hypotension, including those on antihypertensive drug therapy.

As diagnosis and intervention depend on the recorded blood pressure level, it is important that commonly encountered errors be avoided (6). These include falsely high readings, which may result if:

- The cuff is too narrow, too loosely applied, or deflated too slowly.
- The mercury column is read below eye level.
- The subject's arm is below heart level.
- Time has not elapsed before taking the reading of a subject who has been exposed to stress, pain, exertion, or cold.
- An existing arrythmia has not been detected.

Diastole may be read prematurely and/or falsely low systolic pressure may be recorded if the clinician:

- Permits the pressure to drop too quickly (more than 2 mm Hg/sec).
- Fails to identify an auscultatory gap.
- Has difficulty hearing the Korotkoff sounds.

Falsely low systolic readings may also result when:

- The subject's arm is above heart level.
- The mercury column is read above eye level.
- A cardiac irregularity remains undetected.

Falsely low readings delay diagnosis, while falsely high readings may lead to unnecessary concern and the inappropriate initiation of therapy.

In addition to detecting the individual's blood pressure reading, diagnosis in hypertension requires that:

- Any associated signs or symptoms be recorded.
- A search be performed to exclude the presence of any variables that may cause hypertension.
- The subject be screened for evidence of complications attributable to elevated blood pressure.

Hypertension is frequently asymptomatic, hence the screening recommendations. About 45% of persons with hypertension do complain of dizziness; a further 35% complain of headaches. In addition to documenting any clinical manifestations experienced as a result of raised blood pressure, the clinician must identify the presence of any variables that may predispose the patient to hypertension.

The diagnosis of essential hypertension requires exclusion of—while the diagnosis of secondary hypertension is made in the presence of—etiological factors such as renal disease, aortic coarctation, and endocrine-induced and/or drug-induced hypertension (7). Underlying renal disease should be suspected in persons:

- Who are under 30 or over 50 years of age and who present with the sudden onset of hypertension.
- Whose hypertension suddenly worsens.
- Who have hypertension and a history of renal problems.

In patients with hypertension, glomerulonephritis, analgesic nephropathy, and renal trauma can present with hematuria; urinary tract infection and nephrolithiasis with flank pain; and pyelonephritis, vesicoureteral reflux, and urinary tract infection with dysuria and frequency. An abdominal bruit in a hypertensive patient may be indicative of renal artery stenosis. Narrowing of the aorta just distal to the origin of the left subclavian artery presents clinically as coarctation of the aorta. The clinical pattern created by narrowing of the aorta in the region of the ligamentum arteriosum creates a clinical pattern characterized by hypertension in the upper limbs with normal or low blood pressure readings in the lower limbs, weak or delayed femoral pulses, and visible or palpable evidence of collateral circulation in the intercostal spaces or along the borders of the scapulae. Functional clinical findings are exaggerated with exercise. Drugs that aggravate or induce hypertension include steroid preparations (cortisol, androgens, and oral contraceptives), thyroxin and associated preparations, adrenaline, and amphetamines. Endocrine disorders that raise blood pressure include:

- Hyperaldosteronism. The clinical presentation is marked by nocturia, polyuria, muscle cramps, and weakness.
- Pheochromocytoma. The clinical syndrome includes "attacks" of palpitations, headaches, tremors, and, as a result of a hypermetabolic state, excessive sweating and weight loss. Hypertension may be paroxysmal but is frequently sustained. Postural hypotension with associated tachycardia is well documented.
- Cushing's syndrome. The diagnostic pattern of this condition includes truncal obesity associated with peripheral wasting and a moon face; atrophy of the skin with easy bruising, purple striae, and hirsutism; and backache, muscle weakness, osteoporosis, and glycosuria.

In all cases the diagnostic strategy underlying the exclusion of secondary hypertension depends on deductive reasoning and a knowledge of the natural history of the condition ultimately diagnosed. Syndrome or pattern diagnosis provides valuable clues en route to achieving a causal diagnosis with its therapeutic implications. Causal diagnosis excludes all but those cases that are allocated, for want of clearer definition, to the essential hypertension group.

Regardless of the etiological factors involved, once blood pressure is persistently and/or excessively raised above physiologically acceptable limits, complications may occur. Target organs particularly susceptible to the pressure effects of hypertension include the cardiovascular system and the vasculature of the brain, the kidney, and the retina. In hypertensive heart disease, the ventricles expel their contents against increased peripheral resistance. In an effort to maintain a normal cardiac output, the pathophysiological response of the heart is to increase the force of ventricular contraction by myocardial hypertrophy. In cases of systemic hypertension, the thrusting beat of the enlarged left ventricle can be palpated inferolateral to the midclavicular line of the fifth intercostal space. In cases of pulmonary hypertension, the more diffuse pulsation of an enlarged right ventricle can be detected as a parasternal heave. Cardiac auscultation confirms systemic hypertension in that the mitral component of the first and the aortic component of the second heart sounds are accentuated. The left ventricle, as the result of contracting against increased resistance with or without further increases of systemic blood pressure, eventually enters a phase of decompensation. Cardiac failure becomes clinically manifest. The progressive clinical pattern associated with cardiac failure and left ventricular dilation includes:

- Complaints of orthopnea, paroxysmal nocturnal dyspnea, and dyspnea at rest.
- A triple heart rhythm. The fourth heart sound, which occurs with active filling of a dilated ventricle, is heard early in ventricular decompensation;

the third heart sound, which results from passive filling of a dilated ventricle, is a later manifestation of cardiac failure.
- A diastolic blowing murmur indicative of a relative aortic incompetence. In severe cases of hypertension, an early diastolic murmur may be detected at Erb's point as blood under pressure seeps through a "closed" aortic valve.
- Additional confirmatory signs as cardiac failure progresses. These include a pulsus alternans, a regular pulse of varying volume; and crepitations in the lung bases, testimony to a ventricle unable to cope with blood influx causing pulmonary edema.

Hypertension may directly damage the vasculature. Arterial rupture is an acute complication of uncontrolled hypertension. The clinical manifestations depend on the vessel involved. Dissection of the aorta results in profound shock; rupture of a congenital intracranial aneurysm may manifest as the sudden onset of a severe headache and a neurological deficit. Atherosclerosis and the subsequent development of ischemic heart disease are important chronic complications of raised blood pressure. Hypertensive individuals at particular risk for ischemic heart disease are those who carry additional risk factors for atherosclerosis. These include:

- A genetic predisposition, suggested by an individual having parents or siblings diagnosed with ischemic heart disease before the age of 50 years.
- A personal previous diagnosis of familial beta-hyperlipidemia.
- Being a white person, particularly a male or postmenopausal female.
- Increasing age.
- Hypercholesterolemia.
- Smoking.
- Obesity.
- Diabetes mellitus, gout, or hypothyroidism.

Additional evidence of complications attributable to raised blood pressure depends on both the organ involved and the severity of the hypertension. In cases of benign hypertension, cerebral manifestations are limited to headaches and vertigo; in malignant hypertension, cerebral edema, hemorrhage, and/or thrombosis may occur. In the late stages of benign hypertension, renal arteriolar sclerosis may manifest as proteinuria; in malignant hypertension, proteinuria in excess of 2 g/day and renal failure are encountered. Retinal changes associated with hypertension correlate well with the clinical course of the disease and are considered of prognostic significance. Grade 1 hypertensive retinopathy presents with general narrowing of the arterioles described as "silver wiring" of the retinal vessels; in grade 2 retinopathy, arteriolar narrowing is more marked and thickening of the arteriolar walls results in arteriovenous nicking as arterioles cross veins. Grade 3 hypertensive retinopathy demonstrates, in addition to the described vascular changes, flame-shaped and circular hemorrhages and fluffy "cotton-wool" exudates. In benign cases, hypertensive retinopathy is grade 1–3; in malignant hypertension, grade 4 hypertensive retinopathy includes blurring of the disc margins, loss of the physiological cup, and elevation of the optic disc. Papilledema in a patient with elevated blood pressure and other signs of hypertensive retinopathy is diagnostic of malignant hypertension. Hypertension is usually regarded as malignant when it induces severe target organ damage; this is most likely to occur when the diastolic pressure is sustained for long periods at, or rises above, 130 mm Hg.

Detection of raised blood pressure serves as a stimulus to further data collection and decision making. In the absence of an identifiable cause, deductive diagnostic decision making results in labeling the condition essential hypertension. Once the presence of essential hypertension has been established, hypertension as a cause of other diseases is investigated. Probabilistic diagnosis is emphasized in ascertaining the individual's long-term risk of ischemic heart disease; syndrome patterns are sought in determining whether target organs have already sustained damage due to exposure to raised blood pressure levels. The likelihood of rational management decisions being made for patients with essential hypertension is enhanced when the clinician is cognizant of all the relevant diagnostic decision making data.

MANAGEMENT DECISION CRITERIA

Selection of an appropriate treatment regimen for persons with essential hypertension is influenced by a number of factors:

- The patient's blood pressure reading. This is determined on at least three occasions on two separate visits, with due regard for the technical precautions required to justify labeling an individual as hypertensive. The technical demands of accurately assessing blood pressure levels assume trivial proportions when considering the dilemma of demarcating the blood pressure readings that should be regarded as normal and abnormal. The threshold for hypertension may be determined using a statistical approach that designates a point in the blood pressure distribution curve as abnormal; it may, despite the absence of a clear categorical threshold, be demarcated by relating blood pressure levels to the continuously graded curve delineating morbidity and mortality risk; or it may be identified by ascertaining the blood pressure threshold at which the benefits of intervention outweigh the long-term costs and side effects of therapy (8). Depending on the approach favored, normal blood pressure may be regarded as falling below 140/90 mm Hg; borderline hypertensive states as lying between 140/90 and 160/95 mm Hg; and hypertension as existing when blood pressure readings equal or exceed 160/95 mm Hg.
- The patient's age. A diastolic reading of 90 mm Hg in an elderly patient is considered as "healthy" as is a reading of 60 mm Hg in a younger person. Systolic readings of 100 mm Hg are considered "normal" in young people, and 140 mm Hg is "normal" in the elderly.
- The presence of other cardiovascular or atheromatous risk factors.
- A family history of cardiovascular disorders.
- The presence of complications attributable to hypertension.

These data are then subject to the utility approach strategy in which the benefit to be derived from active intervention is weighed against the cost (physical, psychological, and financial) of implementing therapy.

Current thinking maintains that drug therapy is obligatory in all cases of malignant and severe hypertension; it is also indicated in cases of moderate hypertension and should probably be implemented in cases of mild hypertension when other cardiovascular risk factors are present. The implementation of these criteria in chiropractic clinical practice suggests that patients meeting one of the following criteria should be referred:

- Diastolic pressure above 115 mm Hg.
- Diastolic pressure above 90 mm Hg and a demonstrable risk factor for ischemic heart disease.
- Diastolic pressure above 90 mm Hg and one or more of the complications of hypertension.

Patients who do NOT carry any ischemic heart disease risk factor or manifest any complication attributable to hypertension and who fall into the following categories should be reassessed:

- Persons with diastolic pressures above 95 mm Hg should be checked in one month.
- Males under the age of 35 years with systolic pressures above 150 mm Hg should be checked in one month.
- Persons with diastolic pressures of 90–95 mm Hg should be checked within three months.

Nondrug intervention measures may suffice, and immediate referral is therefore not indicated in (9):

- Low-risk persons with diastolic pressure of 90–100 mm Hg.
- Young persons with a diastolic pressure of 85–90 mm Hg and a family history of vascular disease.

Utility reasoning, when applied to clinical intervention in the management of essential hypertension, encounters the problem of adequately defining when the benefits of drug therapy outweigh the costs. While antihypertensive drug therapy has been found effective in reducing the mortality and morbidity attributable to the complications arising from raised blood pressure (10), the beneficial effects of drug therapy in modifying the natural history of mild essential hypertension are disputed. Aggressive drug therapy has been clearly shown to reduce pressure-related complications of hypertension; antihypertensive drug therapy can successfully reduce the risk of accelerated hypertension, stroke, and congestive cardiac failure (11). The effect of drug therapy on the more subtle vascular responses to increased blood pressure levels is less clear. The pharmacological management of malignant hypertension and benign complication hypertension is not disputed, but mild uncomplicated essential hypertension constitutes a persistent therapeutic dilemma.

Although analysis of epidemiological and actuarial information suggests that mild diastolic hypertension and/or mild systolic hypertension adversely affect prognosis, this does not necessarily mean that reducing blood pressure will reverse the situation and favorably affect prognosis (12). While these data suggest that mild hypertension may justifiably be considered an important risk factor for premature cardiovascular mortality, the effect of treatment on ischemic heart disease and other atherosclerotic complications is disputed; certain clinical trials suggest that aggressive drug intervention increases ischemic heart disease mortality. Claims that systematically effective pharmacological care of mildly hypertensive patients improves their prognosis (13) have been challenged by work that queries whether antihypertensive therapy is beneficial (14). Such disparate findings may be attributable to methodological research problems, particularly with regard to patient selection criteria, trial duration, and sample size (15). Findings in a study in which a thiazide diuretic, without a potassium supplement, was used as antihypertensive therapy suggested that the study group had a higher mortality rate than that statistically predicted for a similar population group not on antihypertensive therapy (16). This result may be considered an indictment of the manner in which antihypertensive therapy was prescribed rather than a criticism of the use of antihypertensive agents in mild hypertension. The adverse effect postulated to result from diuretic-induced hypokalemia in these patients has not been proven to the satisfaction of at least one critic (17). A literature analysis led Applegate (18) to conclude that although a significant cardiovascular risk reduction can be achieved by treating mild to moderate hypertension in elderly patients, the magnitude of the absolute risk reduction is not so great that treatment should invariably be pursued if serious side effects cannot be avoided. Refining a diagnosis of essential hypertension to mild essential hypertension further emphasizes the problem of rational management decision making when diagnostic labeling camouflages ignorance about disease etiology.

When faced with contradictory information, the clinician may seek guidance from literature reviews published in refereed journals and from the deliberations of consensus conferences. Some standardization in clinical practice is achieved in conventional health care by implementation of recommendations prepared by expert committees. These expert panels base their conclusions on current scientific evidence. Conventional knowledge in the management of conditions of unknown etiology and debated pathogenesis may be largely influenced by the consensus achieved by experts after consideration of currently available, preferably scientific, information. By 1979, for example, nondrug hypertension management issues that had been subject to consensus conference "resolution" included sodium intake and excess energy intake/obesity (19). Appropriately structured, the consensus process does provide a means of reviewing research and practice information; ideally, consensus conference enunciations may be graded from firm recommendations based on rigorous research evidence to tentative guidelines representing inadequately substantiated belief (20). In the absence of rigorous scientific standards, distortions and delusions are inevitable. The challenging of contradictory results by more research can identify and resolve methodological defects, but consensus conferences may impede such scientific progress (21). Nonetheless, the recognition that consensus conference resolutions are interim guidelines rather than concrete rules does provide a sensible approach in a progressive area.

In December 1985, the 1982 guidelines for the treatment of mild hypertension were revised; the new guidelines are based on the "best available" scientific evidence and will be updated to stay abreast of developments in this field. Currently, the World Health Organization's guidelines for the treatment of mild hypertension, defined as a phase V diastolic blood pressure reading of 90–104 mm Hg in the absence of organ damage, are (22):

- Cessation of smoking and implementation of a nonpharmacological regimen in persons whose diastolic blood pressure remains, on average, below 100 mm Hg. Blood pressure should be monitored on several occasions over the next three months.
- The initiation of drug therapy in persons whose diastolic blood pressure remains, on average, over 99 mm Hg during the three-month monitoring period while on nondrug intervention.
- Reinforcement of nonpharmacological measures and formulation of a long-term monitoring program in individuals whose blood pressure is usually below 100 mm Hg during the three-month observation trial.
- Reconsideration of drug therapy if, after a second three-month nondrug therapy and monitoring period, diastolic pressure levels are 95 mm Hg or greater.
- Reevaluation of any decision to withhold drug therapy in persons who diastolic pressures remain at 90–95 mm Hg and who have an increased risk of cardiovascular disease attributable to smoking, diabetes, hypercholesterolemia, or another of the recognized risk factors.

Additional factors that influence the decision to initiate early drug therapy include a high systolic blood pressure, cardiac or renal disease, and a family history of stroke or cardiovascular pathology. The goal of therapy is to reduce blood pressure to normotensive levels; lowering diastolic pressure to below 90 mm Hg is particularly emphasized.

Clinical trials suggest that nondrug measures are capable of reducing both systolic and diastolic blood pressure levels; the magnitude of the response varies but is frequently of the order of 3–6 mm Hg. This response is sufficient to achieve reclassification of a mildly hypertensive patient. In the absence of target organ disease and/or in the presence of other risk factors for ischemic heart disease, nondrug therapy is recommended for persons with a diastolic blood pressure of 90–94 mm Hg (23). While the success of nondrug intervention measures requires further clinical validation, dietary modification, weight loss, exercise, and relaxation techniques have all achieved some clinical success in the management of mild essential hypertension (24–28). These interventions may also be deemed to have a general health promoting affect. As long as the etiology and pathogenesis of essential hypertension remain obscure, diverse clinical interventions that correlate with improved blood pressure recordings and are apparently devoid of deleterious side effects are justifiable. This is particularly true when the proposed intervention has inherent health promotion value. As long as clinical benefit is validated and side effects are minimal or absent, a favorable cost–benefit ratio can be achieved. When clinical benefit is postulated rather than proven, the potential for a cost–benefit ratio that approximates infinity cannot be ignored.

Despite the "safety" associated with the use of nondrug measures in the control of mild hypertension, certain authorities consider that "there is sufficient evidence regarding the efficacy of most of the nonpharmacological measures in inducing *sustained* blood pressure reduction and no evidence that nonpharmacological treatment reduces hypertension-related morbidity and mortality" (12). Not all agree:

> Recent evidence indicates that nonpharmacologic therapy, including dietary potassium and calcium supplements, reduction of salt intake, weight loss for the obese patient, regular exercise, a diet high in fiber and low in cholesterol and saturated fats, smoking cessation, and moderation of alcohol consumption produces significant sustained reductions in [blood pressure] while reducing overall cardiovascular risk (29).

Concern about the risk–benefit ratio of pharmacological intervention in mild hypertension has led, through a consensus process, to support for evaluating the individual response of persons with mild hypertension to nonpharmacological intervention (30). A step 1 therapy protocol that recommends nondrug intervention has been formulated (31). This protocol, which aims to achieve a diastolic pressure of less than 90 mm Hg, prescribes:

- Exercise, as recommended and approved (26).
- Behavior modification using relaxation, biofeedback, and/or stress reduction (27).
- Dietary modification with reduction in energy intake to ensure a weight within 15% of ideal body weight, a sodium intake of 70–90 mmol/day, and alcohol intake of less than 2 ounces/day. Dietary increases of potassium, calcium, and magnesium, as well as a polyunsaturated:saturated fat ratio above 1.0, may also be considered.
- Decrease or discontinuation of tobacco use.

In a reappraisal of the policy for the prevention and control of hypertension in New Zealand, it has been suggested that the first priority in prevention should be a population strategy directed at encouraging healthy nutritional and exercise habits (32). This population approach to mild essential hypertension was proposed because 57% of excess deaths due to raised blood pressure in New Zealanders occur in persons with a diastolic blood pressure range of 80–94 mm Hg. In view of the potential risks of drug therapy outweighing the benefits of intervention in apparently normotensive and mildly hypertensive persons, only nondeleterious "natural" interventions can be justified. In view of the excess morbidity and mortality associated with mild essential hypertension, an effective form of nonpharmacological intervention is essential. The adrenergic nervous system, renal function, particularly as influenced by hormone-induced sodium loss and/or sodium retention, and the renin–angiotensin system are important regulators of cardiovascular physiology. Antihypertensive drug therapy targets these mediators of pressor control; effective nondrug interventions should also have an impact at one or more of these levels of blood pressure control.

Aerobic exercise is currently recommended as an adjunct in the management of hypertension. While isometric exercise is contraindicated in hypertensive subjects owing to its ability to increase both systolic and diastolic blood pressures (33), controlled isotonic physical activity, despite its ability to elevate systolic pressures in all subjects and diastolic pressure in hypertensive individuals (34), is regarded as a useful therapy. Moderate regular exercise, e.g., cycling at 60–70% of maximum work capacity of 45 minutes three times a week, lowers both systolic and diastolic blood pressure in sedentary persons with mild or moderate hypertension (35). Regular exercise even caused a reduction of 10/7 mm Hg in a control group of normotensive individuals. The mechanism whereby blood pressure falls in response to regular exercise is postulated to result from an overall reduction in sympathetic activity and an increase in peripheral vascular capacity. The blood pressure response to isotonic exercise may serve as a predictor of future hypertension in individuals whose resting blood pressure is considered borderline (34).

Psychological methods of lowering blood pressure, which enjoy varying success depending on the responsiveness of the patient, include psychotherapy, relaxation, and biofeedback. Publications favorable to these techniques report reductions in blood pressure of 20/12 mm Hg following six sessions of relaxation therapy, 9/8 mm Hg after six months of Buddhist meditation, or 10 mm Hg using biofeedback (27). Placebo trials have also shown response, with blood pressure reductions of 5–25 mm Hg. Biochemical studies of thermal biofeedback treatment found that biofeedback-treated patients showed a reduction in mean arterial pressure and norepinephrine levels; no changes were detected in plasma renin activity, plasma aldosterone, and urinary sodium or potassium (36). Trials comparing thermal biofeedback and relaxation training in the management of essential hypertension failed to identify an intervention that was consistently effective in reducing vascular activity across a range of stressors (mental arithmetic, cold stress, negative mental imagery) (37), but scales are being developed that appear to discriminate between persons who will and will not succeed in blood pressure self-regulation (38). Relaxation and stress management seem to be viable clinical interventions in appropriately structured general medical practices (39). Motivated chiropractors may find this a clinically useful intervention in the management of mild essential hypertension. In comparing his review of psychological methods with drugs and dietary interventions, Whyte (27) ranks the effectiveness of intervention after five to 26 weeks as greatest for drugs, followed by weight loss, yoga, salt restriction, meditation, exercise and biofeedback, muscle relaxation, and finally placebo.

Of all postulated dietary factors influencing blood pressure levels, the three that best withstand scientific scrutiny are sodium, alcohol, and excess energy ingestion (obesity). The general belief that salt intake and blood pressure are related is largely supported by epidemiological evidence and partly confirmed by clinical evidence. In their analysis of epidemiological evidence, Morgan and Myers (40) substantiate the salt–hypertension hypothesis by referring to studies showing that:

- In populations with a daily sodium intake of less than 20 mmol (and a potassium intake of 100 mmol), hypertension is virtually absent and blood pressure does not rise with age.

- In the Japanese population the daily sodium intake exceeds 400 mmol/day and the prevalence of hypertension is 50%. (This population also has a low potassium intake.)
- In Australia, where one in five persons has hypertension, the sodium intake in males is 180 mmol/day and in females is 140 mmol/day.
- In an Iranian group of pastoral nomads who have a sodium intake of 150–200 mmol/day, blood pressure rises with age.

These finding suggest that the greater the dietary salt intake, the greater the likelihood of higher blood pressure readings and the advent of hypertension. This hypothesis has been substantiated by clinical trials showing that dietary sodium restriction is accompanied by a drop in blood pressure (41,42). Successful clinical trials suggest that dietary restriction of sodium to about 70 mmol/day will result in a reduction of diastolic blood pressure of the order of 7 mm Hg. Unfortunately, not all clinical trials have similar findings, and those that do support the sodium–hypertension hypothesis often either lack adequate control or are carried out on small samples. A variant on the salt–hypertension hypothesis is proposed by the study of Ambrosioni and associates (43), which found that dietary salt restriction, while not affecting the resting blood pressure level, did reduce the pressor response to stress and exercise. This finding further supports a mechanism proposed to explain the pathogenesis of essential hypertension in the face of increased dietary salt consumption. Abernethy (44) suggests that dietary sodium intake influences the "set point" for the renal pressure–diuresis mechanism and hence blood pressure control. Lower salt intakes will lower the "set point" and higher dietary salt loads raise the "set point," making the system hyperresponsive to pressor stimuli. As the pressure–diuresis mechanism of any individual is genetically controlled, the sensitivity of this mechanism to achieving a "set point" in the face of a continually high sodium load is ultimately genetically determined. A salt-sensitive subgroup may explain some of the discrepancies encountered in clinical trials restricting salt ingestion in hypertensive patients (45). The existence of a sodium-sensitive subgroup would support the use of sodium restriction in certain patients. Identification of sodium-sensitive patients in whom sodium restriction may unreservedly be implemented does, however, constitute a problem. While further investigation is required, potentially suitable candidates for this subgroup may include those with a family history of hypertension (46) and those with a low initial renin value who respond with a smaller rise in renin levels following dietary sodium restriction (47).

The use of salt restriction does provide a potential means of controlling mild essential hypertension, but its use is not without reservation. Doyle's (48) critical review of the clinical benefits of Kempner's rice diet, artificially low sodium diets, and sodium diets with and without diuretics casts doubt on the value of sodium depletion as a means of controlling mild and borderline hypertension. Sodium restriction as a physiologically effective method of blood pressure control may be limited to a genetically sensitive group. Until this is conclusively proven, the patient should not be deprived of a possibly beneficial intervention measure. Persons in whom it may be clinically justifiable to institute sodium restriction include those who are:

- Genetically salt-sensitive.
- Borderline or mildly hypertensive.
- On antihypertensive drug therapy. Sodium restriction has been shown to improve blood pressure control in patients on thiazide diuretics (49).
- Being weaned off antihypertensive drug therapy. Weight loss and/or sodium restriction were found to increase the success rate in a clinical trail in which mildly hypertensive patients, who had been well controlled on one antihypertensive agent, were successfully "unmedicated" (50).

The practical implications of restricting a patient's sodium intake to 70–100 mmol/day include:

- No addition of salt during food preparation. Flavor may be added in the form of garlic, wine, herbs, and spices.
- No addition of salt to prepared food.
- Avoidance of all heavily or visibly salted items.
- Avoidance of most processed foods, e.g., bacon, peanut butter.
- Replacement of salted sauces, pickles, or chutney with vinegar, lemon juice, or dry mustard. "Low" and "no-salt" seasonings are also available (51).

In addition to being found in sodium chloride or table salt, sodium may be ingested in baking powder, monosodium glutamate, or even vitamin tablets, e.g., sodium ascorbate. Patients should therefore scrutinize all labels. On a diet that permits sodium ingestion of 88 mmol/day, the patient may ingest 2 g of sodium, i.e., 5 g of table salt. This is about one teaspoonful. Taste adaptation to salt restriction does occur after about two months of dietary salt limitation (52). It has been found that a diet containing salt levels of less than 0.3 g/day is unpalatable; this is a far cry from the 4–5 g of salt permitted to patients who are on a moderate sodium restriction regimen for the management of essential hypertension.

Salt, i.e., sodium, restriction as a nondrug therapy or adjunctive intervention technique in persons with essential hypertension does appear to have some circumstantial and clinical validity. Most of the major objections to dietary sodium restriction as a strategy for the management of mild essential hypertension can satisfactorily addressed (53):

- Excess dietary sodium has not been shown to cause hypertension. Nonetheless, excess dietary sodium does appear to be a necessary, though not sufficient, participant in the pathogenesis of essential hypertension.
- Sodium restriction has not been shown to be effective in reducing blood pressure. The clinical trials may have failed to adequately identify the sodium-sensitive subgroup.
- Sodium restriction has not been shown to be effective in reducing the cardiovascular risk of hypertension. It is naive to exclude sodium restriction on these grounds without simultaneously questioning the impact of hypertension and/or dietary control of hypercholesterolemia on the mortality and morbidity of ischemic heart disease.
- Sodium restriction may be harmful. Animal experiments rather than human studies support this proposition; the unpalatable nature of a very low-

salt diet and the high sodium intake in modern diets also suggest that this criticism may be unwarranted.

Sodium restriction is impractical and unacceptable. Such a criticism fails to address the issues of compliance with other modes of intervention, including persuading patients to take regular antihypertensive medication with its symptom-producing side effects.

Clinicians who consider that salt and sodium are not critical determinants in the genesis of essential hypertension (54) are probably also persuaded that there is no compelling evidence linking potassium, calcium, and magnesium to hypertension. While there is little to suggest that alterations in cation intake will consistently reduce elevated blood pressure to normal levels in population groups, there is some reason to suppose that cations may alter the blood pressure levels in particular individuals (55). None of these hypotheses has been as well investigated in human subjects as the salt (sodium with or without chloride) hypothesis; all have support from epidemiological investigations and animal experiments. Not one of these theories can be discarded until further clinical evidence has been obtained. Mineral abnormalities, other than excess dietary sodium intake, that bear a postulated link to the pathogenesis of essential hypertension include potassium deficiency or a reduction in the potassium:sodium ratio, inadequate dietary calcium, and/or hypomagnesemia.

An increase in dietary potassium may counteract, to some extent, the deleterious effects of sodium on blood pressure. Certainly, epidemiological studies show that high-potassium, low-sodium diets correlate positively with normotensive populations (40). Modern diets with processed foods show reduced potassium intakes—our daily potassium intake is approximately 30–80 mmol, one-third of that of our ancestors (56). A patient's potassium intake may be improved by recommending increased ingestion of fresh fruit and vegetables. Steaming rather than boiling vegetables not only minimizes flavor loss, it also avoids excess loss of potassium in cooking water.

When 17 nutrients were correlated with the blood pressure profiles of over ten thousand Americans, a low calcium intake was found to be the most consistent factor in hypertensive individuals (57). Low dietary calcium has also been identified as a risk factor in at least eight epidemiological surveys (58). These findings are supported by McCarron's (58) work showing that calcium supplementation will reduce blood pressure in spontaneously hypertensive rats. Three clinical studies—two of which were double blind and placebo controlled—have found that diastolic blood pressure can be reduced with calcium supplementation (59–61). The reduction in diastolic blood pressure on a calcium supplement of 1 g/day is of the order of 3 mm Hg. It has even been suggested that an individual consuming less than 300 mg of calcium per day has an 11–14% risk of developing hypertension; this risk is more than halved in persons on a relatively calcium-rich diet of 1,200 mg/day. Diverse statutory bodies recommend a daily intake of 400–1,100 mg of calcium (62). Despite some evidence supporting the notion that calcium deficiency may be implicated in the genesis of essential hypertension, 80% of family practitioners surveyed in Toronto omitted to recommend that their patients with essential hypertension take a calcium supplement (63). An interesting aside to the calcium–essential hypertension postulate is the suggestion that an adequate dietary sodium intake is required before calcium can exert its full protective effect in the normalization of blood pressure levels (64)! Reduced consumption of magnesium, like calcium, correlates with an increased risk of developing hypertension (65,66). In rats, a dietary deficiency of magnesium has been found to reduce the lumen size of the microvasculature (67); this may in turn lead to an increased peripheral resistance and hypertension.

While the role of these mineral deficiencies in the pathogenesis of essential hypertension requires further validation, it is worth noting that eating fruit, vegetables, and unrefined grains will increase the dietary consumption of these minerals. Vegetarian diets are rich in fruit and vegetables. Epidemiologically, vegetarians have been found to be less at risk for hypertension than the general population (68). Rouse and coworkers (69) tested the pressor effects of a vegetarian diet in a controlled trial on normotensive subjects. After adjustment for diverse variables such as age, obesity, and weight change, the authors concluded that a lacto-ovo-vegetarian dietary change induced a lowering of both systolic (5–6 mm Hg) and diastolic (2–3 mm Hg) blood pressures (69). The dietary constituents responsible for the fall in blood pressure have not been identified. In addition to an increased intake of calcium, magnesium, and potassium and a low intake of sodium, vegetarian diets are rich in polyunsaturated fatty acids and dietary fiber (70).

Dietary fat is a much investigated variable in the pathogenesis of atherosclerosis; it has, until recently, not received special attention in the natural history of essential hypertension. In a controlled, randomized trial, both normotensive and hypertensive subjects demonstrated a drop in systolic and diastolic blood pressures when given a low-fat diet with a high polyunsaturated:saturated (1.0) fat ratio (71). Hypertensive subjects showed a drop of approximately 8 mm Hg in both systolic and diastolic blood pressures. During the switch-back phase of the trial when subjects reverted to their normal diets, their blood pressures reverted to pretrial levels. It is interesting to note that a trial group placed on a low-sodium diet failed to achieve any meaningful blood pressure change during the study. An absolute dietary increase of linolenic acid (1%) has also been linked to a 5 mm Hg reduction in both systolic and diastolic pressures (72). Linolenic acid, an essential polyunsaturated fatty acid, is a precursor of prostaglandins and other vasoregulators.

Alcohol, frequently absent from the vegetarian diet, has been linked to hypertension. Epidemiological evidence shows that alcohol is one of the major dietary variables with a consistent and independent relationship to hypertension (73). At least three randomized controlled clinical trials support the hypothesis that alcohol in doses of 27–80 g/day tends to raise blood pressure (74). It has been suggested that each 10 g of alcohol correlates with a 1/0.5 mm Hg increase in blood pressure; this may only be significant in persons who consume more than 20 g of alcohol each day (75). Men consuming more than 350 ml of alcohol a week (about seven glasses of beer a day) are found to have higher systolic blood pressures than teetotalers, regardless of smoking habits (76). The threshold at which alcohol increases blood pressure levels appears to decrease with age, and to be lower in women. In older men, blood pressure may be raised at alcohol intake levels 20–30 ml/day; in women, increases in blood pressure occur above an alcohol intake of 20 ml/day, compared with the 40 ml/day threshold in men (77). Many mechanisms, including increased sympathetic reactivity, hyperinsulinism with resultant sodium retention, and decreased intracellular magnesium and potassium, have been proposed to explain these findings; none has been subject to adequate experimental/clinical verification (75). Alcohol is postulated to modify blood pressure levels both directly and more indirectly through obesity (78).

Epidemiological and animal experimentation supports the conclusion that weight reduction is accompanied by a fall in blood pressure. Weight loss in obese patients reliably leads to some reduction of blood pressure in hypertensive patients (53, 79). These findings have been reaffirmed by the Chicago Coronary Prevention Evaluation Program, which showed a significant correlation between change in weight and blood pressure (25). These investigators went so far as to suggest that a sizable drop in both systolic and diastolic blood pressures may be sustained by a 6% drop in body weight—even when this weight reduction fails to achieve ideal body weight. As with the other general dietary–hypertension hypotheses, weight reduction as a means of blood pressure control in essential hypertension lacks adequate identification of the dietary constituents responsible for observed pressor changes. Although studies that show a positive correlation between weight loss and blood pressure reduction are frequently confounded by sodium restriction, weight loss is a standard and strongly supported intervention measure in the management of mild essential hypertension.

Persons with hypertension are routinely advised to stop smoking, yet cigarette smoking is not a risk factor for chronic hypertension; this despite the fact that nicotine causes transitory increases in blood pressure and the risk of chronic hypertension accelerating to malignant hypertension is increased in smokers (80). While nicotine has not been substantially implicated in the pathogenesis of sustained essential hypertension, it has been implicated as a risk factor in the genesis of ischemic heart, peripheral vascular, and cerebrovascular diseases (81). As patients with essential hypertension are often asymptomatic, it is pertinent to determine whether an intervention both lowers the blood pressure level and reduces or reverses other cardiovascular risks associated with hypertension (82). The degree of efficacy of an intervention may be better reflected by its ability to modify cardiovascular risk factors than by its reduction in blood pressure levels. While there is no good experimental evidence that smoking cessation reduces the risk of ischemic heart disease, observational data suggest that smokers who quit do have a lower mortality from ischemic heart disease than those who continue smoking (83). It is also currently postulated, with some experimental support, that modifications to diet and smoking habits can contribute to the prevention of cardiovascular disease (84). In addition to its effects on the risk factors associated with hypertension, cigarette smoking has implications for the selection of antihypertensive drugs (85). Cigarette smoking does, in the short term, increase heart rate, arterial blood pressure, and catecholamine levels; it activates the sympathetic nervous system.

In contrast to tobacco, chiropractic adjustment is postulated to normalize raised blood pressure through modification of the tonicity of the autonomic nervous system. Although currently lacking scientific validation as an intervention in the management of essential hypertension, chiropractic adjustment is worthy of further investigation. A controlled clinical trial using a mechanical chiropractic adjusting device showed that hypertensive patients responded favorably to adjustment of subluxations affecting the thoracic spine (86). Although this study found that both diastolic and systolic readings were reduced following thoracic adjustment, the persistence of this normotensive response needs investigation. A technique that produces a transitory lowering of blood pressure does not offer a viable option in the management of essential hypertension. Crawford and coauthors (87) reviewed the current literature and concluded that, on balance, most patients with essential hypertension would benefit from regular chiropractic care. This is contrary to current thinking. A special subcommittee constituted to evaluate routinely prescribed nonpharmacological measures used in the control of essential hypertension concluded that only weight loss, sodium restriction, and limitation of alcohol ingestion have sufficient scientific justification to warrant their recommendation (88).

The management of essential hypertension, regardless of whether it emphasizes a nutritional or pharmacological approach, is beset by the problem of requiring long-term adherence to an intervention program. This problem is exacerbated by the fact that hypertensive patients are often asymptomatic. Attempts to cope with this dilemma include admission of hypertensive patients to special health education programs (89), and the development of patient/client health management contracts (90). Fortunately, contemporary health care is experiencing a change in patient–practitioner relationships (91); patients are assuming increased responsibility for their health care (82). This trend augurs well for a nondrug approach to the control of hypertension, as personal prerogative is an inherent component of management decisions that revolve around behavioral modification.

IN PERSPECTIVE

A number of important issues must be addressed before apparently healthy individuals are labeled as having essential hypertension. Clinical decisions that will profoundly affect the individual's future well-being include:

- Determination of the individual's blood pressure level as normal, borderline, or hypertensive.
- Determining whether drug therapy is obligatory, optional, or even contraindicated. Variables that influence this stage of diagnostic decision making involve consideration of:
  - The current and persistent blood pressure level recorded in the patient.
  - Evidence of target organ damage.

- The presence of risk factors, other than hypertension, that predispose to cardiovascular disease.
- The particular type of intervention selected. Therapy can be modified depending on:
  - The efficacy of the intervention, efficacy being a measure of the ability to both lower raised blood pressure levels and modify cardiovascular risk factors.
  - The dominant mechanism(s) postulated to be involved in the genesis of the hypertension. Clarification of the diverse mechanisms by underlying the pathogenesis of hypertension may lead to physiological markers that can provide management guidelines (93).
  - The side effects best tolerated by the patient. Although the therapeutic success of drugs in moderate and severe hypertension justifies their use, side effects are a problem. Centrally acting antihypertensives such as reserpine are associated with depression, β-blockers with postural hypotension, and diuretics with electrolyte, particularly potassium, imbalance. Excessive lowering of diastolic blood pressure can even result in impaired myocardial perfusion in persons with atheromatous coronary artery changes (94).
  - The likely compliance of the patient. Sociocultural beliefs, health values, and economic considerations all impinge on the likelihood of the patient adhering to any proposed intervention strategy. For chronic conditions in which long-term intervention depends on patient cooperation, the ultimate success or failure of management is probably more a function of patient compliance than of the efficacy of the intervention strategy. The importance of patient cooperation in implementing a mutually agreed upon protocol in the management of asymptomatic essential hypertension cannot be overemphasized.

Patient compliance will remain a challenge in the clinical management of essential hypertension. Identification of clinical markers that will enable practitioners to select the persons most likely to respond to nondrug intervention would be a major contribution to clinical decision making in the management of mild essential hypertension. Development of such markers would be facilitated by an understanding of the genesis of essential hypertension and an analysis of indicators used to define psychosocial compliance. Identification of such markers would largely overcome the ignorance currently camouflaged by the precision of diagnostic labeling in this condition, and would facilitate the implementation of safe clinical care.

REFERENCES

1. Taffinder AP, Taffinder GA: An audit of hypertension in general practice. *Practitioner* 228:595–598, 1984.
2. Prineas RJ, Stephens WB, Lovell RRH: Blood pressure and its treatment in a community. *Med J Aust* 1:5–9, 1973.
3. Simpson FO: *A Guide to the Selection of Patients for Anti-hypertensive Treatment in Blood Pressure Screening in New Zealand.* Christchurch, National Heart Foundation of New Zealand, 1977.
4. Canadian Task Force, The periodic health examination. *Can Med Assoc J* 121:1195–1254, 1979.

5. Hunyor SN, Kewal NK: Screening for high blood pressure—do's and don't's. *Patient Manage* July:91–101, 1985.
6. Raftery EB: Problems in the measurement of blood pressure. *Aust Fam Physician* (Special Issue) 4:48–50, 1975.
7. Jamison JR: *The Practitioner's Handbook of Disease Intervention: The Cardiovascular System.* Melbourne, PIT Press, 1984, pp 21–36.
8. Rocella EJ, Bowler AE, Horan M: Epidemiologic consideration in defining hypertension. *Med Clin North Am* 71:785–801, 1987.
9. Morgan TO: Drug or nondrug therapy for mild hypertension? *Curr Ther* Dec:39–42, 1981.
10. Veterans Administration Cooperative Study Group on Antihypertensive Agents. Effects of treatment in morbidity in hypertension. *JAMA* 202:116–122, 1967.
11. Black ER: Treatment and control of hypertension as preventive therapy. *Primary Care* 16:49–62, 1989.
12. Cressman MD, Vlasses PH: Recent issues in antihypertensive drug therapy. *Med Clin North Am* 72:373–397, 1988.
13. Hypertension Detection and Follow-up Program Cooperative Group. Five year findings of the hypertension and follow-up program. *JAMA* 242:2562–2571, 1979.
14. Morgan TO, Adams WR, et al.: Failure of therapy to improve prognosis in elderly males with hypertension. *Med J Aust* 2:27–31, 1980.
15. O'Kelly BF, Massie BM, Tubau JF, Szlachcic J: Coronary morbidity and mortality, pre-existing silent coronary artery disease and mild hypertension. *Ann Intern Med* 110:1017–1026, 1989.
16. Multiple Risk Factor Intervention Trial Research Group. Multiple risk factor intervention trial. *JAMA* 248:1465–1477, 1982.
17. Moser M: Diuretics in the management of hypertension. *Med Clin North Am* 71:935–946, 1987.
18. Applegate WB: Hypertension in elderly patients. *Ann Intern Med* 110:901–915, 1989.
19. Ahrens EH: Introduction to the report of the task force on the evidence of diet related diseases. *Am J Clin Nutr* 32:2627–2631, 1979.
20. Lomas J, Anderson G, Murray E, Vayda E, Roberts R, MacKinnon B: The role of evidence in the consensus process. *JAMA* 259:3001–3005, 1988.
21. Feinstein AR: Fraud, distortion, delusion, and consensus: the problems of human and natural deception in epidemiological science. *Am J Med* 84:475–478, 1988.
22. WHO Statement. Guidelines for the treatment of mild hypertension: 1986. *Med J Aust* 144:575–578, 1986.
23. Gifford RW: Mild hypertension: critical analysis of different therapeutic approaches. *Cleve Clin J Med* 56:337–345, 1989.
24. Rouse IL, Beilin LJ: Nutrition, blood pressure, and hypertension: a critical review of dietary intervention studies in humans. *Med J Aust* 2:S19–S23, 1983.
25. Stamler J, Farinaro E, et al.: Prevention and control of hypertension by nutritional-hygienic means. *JAMA* 243:1819–1823, 1980.
26. Dwyer T, Bruggs DA: The role of physical activity. *Med J Aust* 2:S9–S12, 1983.
27. Whyte HN: Psychological methods of lowering blood pressure. *Med J Aust* 2:S13–S16, 1983.
28. Kaplan NM: Non-drug treatment for hypertension. *Ann Intern Med* 102:359–373, 1985.
29. Oparil S, Calhoun DA: Hypertension. *Dis Mon* 35:133–232, 1989.
30. Horan MJ, Rocella EJ: Nonpharmacologic treatment of hypertension in the United States. *Health Psychol* 7(suppl):267–282, 1988.
31. Saunders E: Hypertension in blacks. *Med Clin North Am* 71:1013–1043, 1987.
32. Beaglehole R, Bonita R, Jackson R, Stewart A: Prevention and control of hypertension in New Zealand: a reappraisal. *N Z Med J* 101:480–483, 1988.
33. Stopford JL: Static exercise—physiologic dangers and proper training techniques. *Nurse Pract* 13:10–18, 1988.
34. Palatani P: Blood pressure behaviour during physical exercise. *Sports Med* 5:353–374, 1988.
35. Nelson L, Jennings GL, Elser MD, Korner PI: Effect of changing levels of physical activity on blood pressure and haemodynamics in essential hypertension. *Lancet* 2:473–476, 1986.
36. McCoy GC, Fein S, Blanchard EB, Wittrock DA, McCaffrey RJ, Pangburn L: End organ

changes associated with the self-regulatory treatment of mild essential hypertension? *Biofeedback Self Regul* 13:39–46, 1988.
37. Blanchard EB, McCoy GC, Wittrock D, Musso A, Gerardi RJ, Pangburn L: A controlled comparison of thermal biofeedback and relaxation training in the treatment of essential hypertension. II. Effects on cardiovascular reactivity. *Health Psychol* 7:19–33, 1988.
38. Nakagawa KH, Garber A, Jarrett M, Egan KJ, Hendershot S: Self-management of hypertension: predictors of success in diastolic blood pressure reduction. *Res Nurs Health* 11:105–115, 1988.
39. Patel C, Marmot M: Can general practitioners use training in relaxation and management of stress to reduce mild hypertension? *Br Med J (Clin Res)* 296:21–24, 1988.
40. Morgan T, Myers J: Dietary salt and hypertension. *Aust Fam Physician* 11:264–267, 1982.
41. Morgan T, Myers J: Hypertension treated by sodium restriction. *Med J Aust* 2:396–397, 1981.
42. Parijs J, Joosens JV, et al.: Moderate sodium restriction and diuretics in the treatment of hypertension. *Am Heart J* 85:22–34, 1973.
43. Ambrosioni E, Costa FV, et al.: Effects of moderate salt restriction on intralymphocytic sodium and pressor response to stress in borderline hypertension. *Hypertension* 4:789–794, 1982.
44. Abernethy JD: Patient compliance in hypertension: role of diet. *Patient Manage* June:23–30, 1977.
45. Ando K, Fujita T. Abnormal renal haemodynamics in salt-sensitive patients with essential hypertension. *Jpn Circ J* 49:984–989, 1985.
46. Skrabal F, Hamberger L, Cerny E: Salt sensitivity in normotensives with and salt resistance in normotensives without heredity of hypertension. *Scand J Clin Lab Invest* 176(suppl): 47–57, 1985.
47. Morgan TO, Teow BH, et al.: Clinical effects of salt restriction on essential hypertension. *Proc Nutr Soc Aust* 9:20–26, 1984.
48. Doyle AE: Sodium restriction. *Med J Aust* 2:S17–S18, 1983.
49. Carney S, Morgan T, et al.: Sodium restriction and thiazide diuretics in the treatment of hypertension. *Med J Aust* 1:803–807, 1975.
50. Langford HG, Blaufox MD, et al.: Dietary therapy slows the return of hypertension after stopping prolonged medication. *JAMA* 253:657–664, 1985.
51. Greenfiels H, McCullum D, Wills RBH: Sodium and potassium contents of salts, salt substitutes, and other seasonings. *Med J Aust* 140:460–462, 1984.
52. Beauchamp GK, Bertino M, Engelman K: Modification of salt taste. *Ann Intern Med* 98:763–769, 1983.
53. Kaplan NM: Nonpharmacologic therapy for hypertension. *Med Clin North Am* 71:921–932, 1987.
54. Dwyer T: Salt and hypertension. *Med J Aust* 142:497–498, 1985.
55. Maxwell MH, Waks AU: Cations and hypertension: sodium, potassium, calcium and magnesium. *Med Clin North Am* 71:859–875, 1987.
56. MacGregor GA: Dietary sodium and potassium intake and blood pressure. *Lancet* 1:750–752, 1983.
57. McCarron DA, Morris CD, et al.: Blood pressure and nutrient intake in the United States. *Science* 224:1392–1398, 1983.
58. McCarron DA: Is calcium more important than sodium in the pathogenesis of essential hypertension? *Hypertension* 7:607–627, 1986.
59. Belizan JM, Villar J, et al.: Reduction of blood pressure by calcium supplementation in young adults. *JAMA* 249:1161–1165, 1983.
60. Grobbee DE, Hofman A: Effect of calcium supplementation on diastolic blood pressure in young people with mild hypertension. *Lancet* 2:703–707, 1986.
61. McCarron DA, Morris CD: Blood pressure response to oral calcium in persons with mild to moderate hypertension. *Ann Intern Med* 103:825–831, 1985.
62. International Union of Nutritional Sciences Committee. Recommended dietary intakes and allowances around the world: an introduction. *Food Nutr Bull* 4:34–45, 1982.
63. Hayman EK: Dietary calcium: a low priority in treatment of essential hypertension by Toronto family practitioners. *J Can Chiropractic Assoc* 30:25–28, 1986.
64. Parrott-Garcia A, McCarron DA: Calcium and hypertension. *Nutr Rev* 42:205–213, 1984.

65. McCarron DA: Calcium and magnesium nutrition in human hypertension. *Ann Intern Med* 98:800–805, 1983.
66. Magnesium deficiency and hypertension. *Nutr Rev* 42:235–236, 1984.
67. Altura BM, Altura BT, et al.: Magnesium deficiency and hypertension: correlation between magnesium deficient diets and microcirculatory changes in-situ. *Science* 223:1315–1317, 1984.
67. Warnock F, Rouse I: Vegetarian diet. *Aust Fam Physician* 18:1550–1555, 1987.
68. Rouse IL, Beilin LJ, et al.: Blood-pressure-lowering effect of a vegetarian diet: controlled trial in normotensive subjects. *Lancet* 1:5–10, 1983.
70. Rouse IL, Beilin LJ: Nutrition, blood pressure and hypertension. *Med J Aust* 2:S19–S23, 1983.
71. Puska P, Iacono JM, et al.: Controlled, randomized trial of the effect of dietary fat on blood pressure. *Lancet* 1:1–5, 1983.
72. Berry EM, Hirsch J: Does dietary linolenic acid influence blood pressure? *Am J Clin Nutr* 44:336–340, 1986.
73. Harlan WR, Hull AL, et al.: Blood pressure and nutrition in adults. *Am J Epidemiol* 120:17–28, 1984.
74. Burr ML: A tot a day keeps disease away. *Proc Nutr Soc* 47:129–133, 1988.
75. Mathews JD: Is alcohol a preventable cause of hypertension? *Med J Aust* 2:S3–S6, 1983.
76. Arkwright PD, Beilin LJ, et al.: Influence of alcohol, life-style and psychological factors on blood pressure and the incidence of hypertension in a working population. *Annual Report of the Raine Centre for the Study of Epidemiology, Aetiology and Community Control of Hypertension and Atherosclerotic Vascular Disease.* 1980, pp 10–16.
77. Criqui MH: The roles of alcohol in the epidemiology of cardiovascular disease. *Acta Med Scand* 717:73S–85S, 1987.
78. Kornhuber HH, Lisson G, Suschka-Sauermann R: Alcohol and obesity: a look at high blood pressure and stroke. *Eur Arch Psychiatry Neurol Sci* 234:357–362, 1985.
79. Nestel PJ: Obesity and hypertension. *Med J Aust* 2:S6–S8, 1983.
80. Benowitz NL: Pharmacologic aspects of cigarette smoking and nicotine addiction. *N Engl J Med* 319:1218–1230, 1988.
81. Hay DR: Health risks of smoking. *Patient Manage* 11:35–47, 1987.
82. Ball SG: Future approaches to the treatment of hypertension in the light of results of recent trials. *Drugs* 35:S9–S15, 1988.
83. Wood DA: Aetiology of ischaemic heart disease. *Br J Anaesth* 61:3–10, 1988.
84. Stamler J: Risk factor modification trials: complications for the elderly. *Eur Heart J* 9:9S–53S, 1988.
85. Trap-Jensen J: Effects of smoking on the heart and peripheral circulation. *Am Heart J* 115:263–267, 1988.
86. Yates RG, Lamping DL, Abram NL, Wright C: Effects of chiropractic treatment on blood pressure and anxiety: a randomized controlled trial. *J Manipulative Physiol Ther* 11:484–488, 1988.
87. Crawford JP, Hickson GS, Wiles MR: The management of hypertensive disease: a review of spinal manipulation and the efficacy of conservative therapy. *J Manipulative Physiol Ther* 9:27–32, 1986.
88. Subcommittee on Nonpharmacological Therapy. Nonpharmacolgial approaches to the control of high blood pressure. *Hypertension* 8:444–467, 1986.
89. Watson DS: Health education in hypertensive patients. *Aust Fam Physician* 8:315–318, 1979.
90. Jamison JR: *Health Pursuit—the Client's Handbook of Health Promotion.* Melbourne, PIT Press, 1985.
91. Haug MR, Lavin B: Practitioner or patient—who's in charge? *J Health Soc Behav* 22:212–219, 1981.
92. Morris CM: Self-care. *Am J Nurs* March:486–489, 1979.
93. Davidman M, Opsahl J: Mechanisms of elevated blood pressure in human essential hypertension. *Med Clin North Am* 68:301–318, 1984.
94. Cruickshank JM, Thorp JM, Zacharias FJ: Benefits and potential harm of lowering high blood pressure. *Lancet* 1:581–584, 1987.

6

Safe Health Care: The Product of Optimal Clinical Decision Making

A universal health care ethic is that the practitioner should do no harm. A fundamental objective of health care is that the patient's health status be positively influenced. The minimum requirement of any clinical encounter is that it be safe. Safe practice results from the interaction of multiple variables at the patient–practitioner interface. Three of these issues will be addressed in this chapter: safe diagnosis, applying the minimax theory in differential diagnosis; safe therapy, using contraindications as a therapeutic guide; and "safe" knowledge: practice can be no safer than the knowledge base from which it is conducted.

SAFE DIAGNOSIS

The reliability of a diagnostic decision may be enhanced by the sequential use of sensitive and specific tests. In the long term, arrival at an accurate diagnosis is believed to benefit the patient; even in the absence of specific therapy, "knowing the score" is deemed of some value. In the short term, the time taken in precisely evaluating a patient may be detrimental to the patient's health—detrimental in the sense that disease progression continues uninterrupted while the patient's condition is under professional review. The minimax theory takes cognizance of this dilemma and proposes that, given a particular presenting complaint, the list of possible diagnoses be ranked according to worst possible outcome. Consequently, the initial working hypothesis to be evaluated is not the most likely diagnosis based on prevalence data, but the condition with the most unacceptable outcome. The conditions first excluded are those disorders that carry the most serious penalty should they be missed.

Headaches affect at least nine in ten people at some time during their lives.

Most headaches are attributable to sustained muscular contraction due to psychosocial problems, chronic myositis, cervical osteoarthritis, or temporomandibular joint dysfunction; about one in ten are due to vascular causes, most of which are functional in nature (1). Only about two in every 100 cases of headaches are attributable to inflammatory lesions or intracranial masses. From a clinical perspective, the vast majority of headaches are benign; some, however, have a serious prognosis. The minimax theory requires the elimination of a serious, albeit rare, disease before attempting to precisely label the more likely functional condition.

Findings suggestive of serious intracranial disease include (2):

- A severe headache in a patient with no prior history of headaches. Particularly ominous is the severe headache occurring for the first time in an individual over 35 years of age.
- A sudden change in a chronic headache.
- Aggravation of a headache on straining at stool or coughing.
- A persistent, highly localized pain.
- A headache that is present on waking.
- A headache that persists without respite for longer than 24 hours.
- A headache in an individual whose neck is not completely supple (check for a fever), who has a convulsion or any impairment of consciousness.
- A headache in a drowsy, confused patient or one with a memory disturbance or personality change.
- A headache that progressively worsens or persists in a patient who looks ill.

Intracranial lesions may cause local irritation that presents clinically as epilepsy or local pressure effects as suggested by unilateral impairment of neurological function. Such neurological deficits are helpful both in clinching a diagnosis of serious organic disease and in anatomically localizing the lesion. Signs and symptoms of a general increase in intracranial pressure may manifest as:

- A headache that starts unilaterally and later becomes diffuse. The headache is throbbing or bursting in quality, worst on waking, and aggravated by exertion, bending, lying down, or sneezing.
- Projectile vomiting, usually unassociated with nausea.
- Papilledema. Retinoscopy reveals retinal vein congestion, increased pinkness of the optic disc, blurring of the disc margins, obliteration of the physiological cup, and lateral elevation of the optic disc. Hemorrhages in and around the disc may also be noted. Visual acuity is usually preserved.
- Brainstem compression. With a progressive increase in intracranial pressure, evidence of brainstem compression may be manifested by loss of consciousness and coma, extensor plantar reflexes, and dilated pupils that fail to respond to light.

The causes of ominous severe headaches include (3,4):

- A subarachnoid hemorrhage. The leakage of blood from a berry aneurysm may precede rupture by days or weeks in about one in 25 patients. Suggestive findings include a severe headache, often initiated by exertion; minimal blunting of consciousness; vomiting; and, if meningeal irritation is

present, neck stiffness. Applying the principles of the minimax theory to intervention, chiropractors are urged to refer all patients who meet the above criteria. An oozing aneurysm is least likely:
- In a young patient.
- When the headache did not start abruptly or during exertion.
- When repeated examinations have failed to detect any neurological deficit, neck stiffness, or fever.
- In the absence of fever.
- Meningitis. In the early stages, the major differential diagnosis may be influenza. Patients may have a severe headache and fever. Careful examination and monitoring for any evidence of neck stiffness is required.
- Tumors. Tumors may present with increased intracranial pressure manifested as gradually progressive headaches that are worst on waking and straining; focal neurological signs may be helpful.
- Hypertension. When the diastolic pressure exceeds 120 mm Hg, malignant hypertension and hypertensive encephalopathy may be encountered. Underlying causes that require urgent investigation include pheochromocytoma and drug therapy; patients on monoamine oxidase inhibitors who consume tyramine-rich foods or take sympathomimetic cold remedies may precipitate a hypertensive attack.
- Vasculopathies. Temporal arteritis may present with jaw claudication, local tenderness, focal headache, impaired scalp artery pulsation on the affected side, an elevated erythrocyte sedimentation rate, and signs of general systemic illness. The disease is more common after 50 years of age. Spontaneous or posttraumatic dissection of the cranial or cervical artery manifests as a localized headache, Horner's syndrome and tenderness over the carotid, or a vertebral artery bruit. Thrombosis, hemorrhage, and/or embolization of a thrombus or necrotic atheromatous plaque are all causes of a headache associated with varying degrees of neurological deficit, depending on the site of cerebral infarction.
- Acute purulent sinusitis. Headaches localized to the frontal or vertex regions in persons with a nasal or postnasal drip and tenderness over the sinuses should be referred for antibiotic therapy.

It has been suggested that most patients with headaches that are severe, constant, unusual, or associated with neurological signs should have a computed tomographic scan. If this recommendation were implemented in primary practice for each individual with a new severe headache, the economic cost would be exorbitant and the yield small (5). In primary practice, clinical exclusion of a new severe headache remains the dominant strategy. Once serious organic headaches have been excluded, then prevalence assumes importance as the dominant diagnostic guideline. The headaches most frequently encountered in primary practice are muscle contraction and vascular headaches.

Functional vascular headaches are usually attributable either to histamine, as in cluster headaches, or to vasodilation following vasospasm, as in migraine. A combination headache is found when a muscular contraction headache occurs in conjunction with migraine or a cluster headache.

Cluster headaches are severe unilateral periorbital headaches with an abrupt onset and termination (6). Pain lasts for ten to 120 minutes and recurs several times over a 24 hour period. Bouts of headache occur daily for weeks or months. Lacrimation and nasal stuffiness are common; miosis and ptosis may occur. Compared with migraine, cluster headaches occur more frequently (six a day), are of shorter duration (less than ten minutes), and have no prodromata. A family history is only occasionally present; blurred vision, nausea, and vomiting are seldom present.

In migraine, a positive family history, nausea, and vomiting are usual; blurred vision is a frequent finding (7). Common migraine, a unilateral throbbing bimonthly headache of four to six hours duration, is relieved by sleep and precipitated by glare, cocoa, or tyramine. Classical migraine has all the features of common migraine plus sensory, motor, or visual prodromata. Prodromata include parasthesia, weakness, scintillating spots, and/or visual field defects. Other variants on common migraine include ophthalmic migraine, in which an aura of bright light is associated with visual symptoms and little or no headache; facial migraine, in which pain starts at the angle of the mouth and spreads to involve the lower half of one side of the face; facial hemiplegic migraine, in which unilateral bouts of weakness and numbness are experienced; opthalmoplegic migraine with the diplopia and ptosis of a third nerve palsy; and vertebrobasilar migraine with a prodrome of vertigo, visual field impairment, tinnitus, and dysarthria. Any headache diagnosed as migraine requires further investigation to exclude a structural lesion if:

- Hemicrania occurs on the same side in each attack.
- Visual or cerebral symptoms are on the side contralateral to the hemicrania.
- Bruits are detected on auscultation of the skull and orbits.
- Any neurological sign persists or another abnormal physical sign is detected.

Management of vascular headaches depends on the cause. Diet has been variously linked with vascular headache (8). Direct-acting vasoactive substances include monosodium glutamate, a flavor enhancer used particularly in Chinese cooking; nitrates, a preservative in cured meats; and alcohol. Nitrates cause facial flushing and a dull headache in sensitive persons. Alcohol has both central and peripheral vasodilatory effects; it is a recognized trigger of both migraine and cluster headaches. In addition to direct chemically induced vasodilation, mechanisms that have been investigated to explain the initiation of migraine include food allergies, hypoglycemia, and taste aversion (9). An established attack of migraine requires drug intervention (10), but avoidance of aggravating or trigger factors and psychological adjustment are recognized components in the management of migraine (11).

The other headache most often seen in primary practice is attributed to muscle contraction. Muscle contraction headaches may be primary or secondary (1,12). Ischemia is believed to play a dominant role in the pathogenesis of this category of headache. Muscle spasm lasting for more than two minutes occludes local blood vessels for a sufficiently prolonged period to precipitate pain. Pain that persists after muscle contraction has relaxed may be referred to the neck, scalp, face, and frontal regions. A headache involving moderate pain lasting for some hours or days and periodically recurring with similar intensity should be labeled a muscle contraction headache. Pain description varies from that of a dull, poorly defined persistent ache to discomfort reminiscent of a tight hatband. Once a working diagnosis of muscular contraction headache has been postulated, it is customary to determine whether the headaches are secondary to some underlying dysfunction. This approach emphasizes the importance of identifying the cause of a condition and targeting therapy at the source rather than at the symptom.

In cases of secondary muscle contraction headaches, the symptomatic condition may result from underlying:

- Cervical discogenic anomalies.
- Cervical osteoarthritis.
- Malocclusion of the temporomandibular joint.
- Chronic mastoiditis.
- A posterior fossa lesion.
- Eye strain.

Patients with muscle contraction headaches attributable to underlying craniocervical lesions present with the following characteristics:

- Ill-defined pain usually described as a dull, nagging ache in the occipital, semicranial, or frontal region. Classically, secondary muscle contraction headaches due to cervical spine dysfunction start at the nape of the neck, spread up to the occiput, and extend forward to behind the eyes.
- Limitation of movement.
- Stiffness and difficulty in initiating movements; this can be improved with exercise.
- Parasthesias.
- Postures requiring prolonged bending of the head anterior to the center of gravity initiate or aggravate the pain.
- A hot water bottle applied to the neck, or holding the head toward the shoulder opposite to the painful side, provides a measure of relief.
- A history of trauma to the head or neck.
- Examination findings demonstrate restricted movement and neurological changes (motor, sensory, and/or reflex). Motion palpation provides particularly important information; the diagnosis may be further substantiated on x-rays of the cervical spine.
- Associated findings range from vertigo, tinnitus, visual problems, and dysphagia to fatigue, irritability, and depression.

Major factors differentiating cervical causes of muscular contraction headaches from tension headaches are an improvement in the headache with movement as the day progresses, the anatomical definition of parasthesias, a history of trauma, and cervical stiffness or grating on motion palpation.

Categorization of secondary muscular contraction headaches is the prerequisite to precise diagnostic labeling. The more precise the diagnostic label, the more relevant the postulated prognosis and the more specific the selected intervention strategy.

Primary muscle contraction or tension headaches are generally believed to be attributable to psychosocial stress. Certainly, such patients are often tense, anxious, or depressed. The classical clinical manifestations of tension headaches include:

- Initial involvement of the occipital region with subsequent spread of pain to the parietal, frontal, and facial regions.
- Pain characteristically described as a squeezing tightness, a constriction, or tense hatband. Pain is bilateral in 90% of cases.
- A sustained or persistent headache of varying intensity.
- Pain that is often relieved by alcohol.
- A normal sleep pattern.
- Headaches that are absent on waking.
- Headaches that develop as the day progresses and daily tensions are encountered.
- The implication that cold draughts are a frequent aggravating and/or precipitating factor.
- Anxiety frequently associated with hyperventilation, resulting in lightheadedness and dizziness.
- The absence of prodromata.
- Tautness of the jaw, neck, scalp, and face musculature on examination.
- The presence of trigger points and tender nodules.

All forms of muscle contraction headache respond to work on soft tissues and improved posture. The definitive intervention for primary muscle contraction headaches includes muscle relaxation training and improved management of psychosocial stresses. Muscle contraction headaches attributable to cervical spine dysfunction respond particularly well to chiropractic adjustment. Table 6.1 compares the clinical presentation of some important headaches that may be encountered in primary practice (13–15).

Early distinctions between muscle contraction headache and migraine are being contested as evidence accumulates documenting psychogenic, autonomic dysfunctional, and biochemical similarities. The positive therapeutic effect achieved by chiropractic manipulative therapy in the treatment of benign headaches may be explained by the hypothesis that this intervention modulates the function of the autonomic nervous system. In any event, somatic dysfunctions associated with headaches—whether they are predominantly attributable to vertebral subluxation, regional myofasciitis, or vertebrogenic autonomic dysfunction—do respond well to chiropractic clinical care (16). With increasing therapeutic validation for the role of chiropractic clinical care in the management of headaches, it becomes particularly important that the minimax theory be applied in the initial diagnostic work up of patients complaining of headaches.

Chest pain is another area in which conditions amenable to chiropractic care may mimic serious organic disease. Musculoskeletal causes of chest pain are usually differentiated from visceral causes on finding local tenderness when palpating involved musculoskeletal structures or restriction of move-

Table 6.1
Differentiating headaches in primary practice (13–15)

	Organic intracranial disorder	Muscle contraction	Migraine	Cluster
Hereditary	No	No	Yes	No
Male:female	—	1:1	1:3	4:1–20:1
Onset (age)	>30 yr	>13 yr	10–40 yr	>30 yr
Frequency	Varies	Daily	1–8/month	1–3/day
Duration	Persists	Hours/days	<12 hr	30–120 min
Intensity	Deep, dull ache	Bandlike ache	Moderately severe	Excruciating
Peak	On waking	Steady	30–60 min	2–5 min
Location	Central	Frontal; bilateral occipital	Hemicranial; temporal	Unilateral oculotemporal
Horner's syndrome	Possible	—	—	+
Aggravating factors	Straining	Stress	Foods, fatigue, noise, alcohol	Alcohol, smoking
Prodromata	—	Absent	Present	Rare
Associated findings	Papilledema	Sore neck	Prodrome nausea	Lacrimation, rhinorrhea
Time when attacks worst	On waking	Afternoon	Varies	Night

ment and exacerbation of symptoms by movement. Musculoskeletal chest pain may arise from the spine, other bones and joints of the thoracic cage, and muscles in the area.

Anterior chest pain produced by cervical lesions may be bilateral, and its characteristics can vary with exertion, respiration, and change of posture (17). The important diagnostic features that help to differentiate between chest pain of spinal and visceral origin are the reproduction of pain on appropriate spinal movement and local joint signs. The underlying pathology may be a hypomobility syndrome, a disc lesion, or vertebral disease. In hypomobility syndrome, the pain may be of sudden onset and exacerbated by movement, breathing, and exertion; pain, usually confined to the paravertebral area, may radiate a short distance along or circumscribe the chest. Pain may occasionally pass directly through the chest. Diagnosis is made on finding a limited range of motion; passive movement may reproduce the pain. Disc lesions and vertebral disease are less easy to confuse with myocardial ischemia. In disc lesions, radicular pain is common, is exacerbated by movement and on lying down, and is associated with sensory disturbance. Vertebral disease, e.g., osteoporosis, multiple myeloma, and other primary or metastatic neoplasms, may present with localized pain and dermatomal radiation. Shooting radicular pain may be confused by pleuritic pain; the stabbing pain of pleurisy is aggravated by deep breathing, which involves movement of the thoracic cage. Involvement of the thoracic cage in the genesis of chest pain can usually be easily identified by local tenderness and/or swelling, restriction of movement, and production of pain on movement.

Fibromyalgia may be a cause of chronic diffuse aching chest pain with trigger point tenderness in the region just lateral to the second costochondral junction (18). As chiropractors have the potential to make a major contribution to the management of fibromyalgia and are therefore likely to retain such patients within their practice for prolonged periods, it is important that stringent attention is paid to eliciting the requisite diagnosis criteria. A diagnosis of fibromyalgia involves identification of two major criteria: chronic diffuse musculoskeletal aches and pains and stiffness of at least three months duration, and at least three trigger points involving more than one location. Important trigger point sites include: the cervical spine, anterior aspects of intertransverse spaces at C5–7; gluteus medius, the upper outer quadrants of the buttocks; 2 cm posterior to the greater trochanter; the medial fat pad of the knee overlying the medial collateral ligament; 2 cm distal to the lateral epicondyle (a tennis elbow site); below the occipital prominences; the midpoint of the upper fold of trapezius; and/or near the medial border of the supraspinatus. Minor criteria include chronic fatigue, emotional distress, poor sleep, and morning stiffness.

While chiropractic management is the treatment of choice for hypomobility syndrome and fibromyalgia, it is contraindicated in myocardial ischemia and pleuritic chest pain. Application of the minimax theory to patients complaining of chest pain requires exclusion of myocardial ischemia, pulmonary embolism, pneumothorax, and pneumonia.

Myocardial ischemia may present as (19):

- Angina pectoris. Pain persists for a few minutes and is associated with physical or emotion stress. Pain is rapidly relieved by rest and/or vasodilator drugs.
- Coronary insufficiency. Pain lasts for several minutes to one hour and may be encountered during rest or on exertion. Pain is incompletely relieved by vasodilator drugs.
- Myocardial infarction. Pain invariably lasts for at least ten to 15 minutes and usually for more than an hour. Pain may occur at any time, usually at rest. Nausea, vomiting, and intense fear are common. Pain is not relieved by vasodilator drugs.

The typical pain of myocardial ischemia is experienced in the center of the chest and is described as burning, squeezing, bursting, or tight. Pain often radiates to the throat or lower jaw, down one or both arms, or to the back. The likelihood ratio of a patient with symptoms of typical angina having at least a 75% coronary artery stenosis (as revealed on angiography or at autopsy) is 115 in men and 120 in women; the likelihood ratio of a patient with a history of a typical angina symptoms having an equivalent loss of coronary artery patency is 14 in men and 15 in women (20). Atypical presentations include pain limited to one or more of the radiation sites or confined to the left mammary area. Pain may rarely be characterized as a "breathlessness." It has been stated that cardiac pain is never transient or well localized and never has a pleuritic component (21). Definitive diagnosis is by serial cardiac enzyme estimations and ECG. Depending on the clinical findings, a resting or exercise ECG may be indicated. An exercise ECG has a sensitivity rate of 60% and specificity quotient of 91%. When creatine phosphokinase, the most frequently assessed cardiac enzyme, is found at levels in excess of 79 IU, the test sensitivity rate is 93% and specificity quotient is 88%. Translated into clinical practice terms, the positive predictive value for myocardial infarction of a creatine phosphokinase test result in excess of 79 IU is 93% in coronary care units; in general hospital admissions this falls to 46%. The positive predictive value of this test in chiropractic clinical practice is not known (22).

Acute respiratory events such as a massive pulmonary embolus or spontaneous pneumothorax are more likely confused with myocardial disease than treated as musculoskeletal disorders (23,24). Massive pulmonary embolism presents with severe persistent retrosternal chest pain resembling myocardial ischemia; smaller emboli may lead to peripheral lung infarction with pleuritic-type chest pain. Diagnosis includes a search for the origin of the embolus; calf veins are suspect. Hemoptysis may be found. Chest pain attributable to spontaneous pneumothorax is initially acute and later pleuritic in nature. A tympanic percussion note, decreased air entry on auscultation, and tracheal deviation are highly suggestive. The diagnosis is confirmed on radiology. Pleuritic chest pain associated with a resolving pulmonary infarction, pneumothorax, or other underlying pathology is differentiated from chest wall pain: the latter is constant and aching and often unrelated to breathing. Detection of underlying lung pathology on percussion, auscultation, and radiology is also helpful in appropriately categorizing sharp stabbing pleuritic chest pain.

Regardless of whether the patient is complaining of headache or chest pain, pain-causing conditions that carry a serious prognosis should be carefully excluded before pain is attributed to functional and/or biomechanical causes. The minimax theory contributes to safe clinical practice by advocating the early and active exclusion of conditions with a poor prognosis. It is one mechanism whereby diagnostic reasoning may contribute to safe clinical practice.

SAFE MANAGEMENT

The goal of diagnostic decision making is to label the presenting complaint of a particular patient according to the current nosological framework. Once collected data suggest that the condition of the patient fulfills the prerequisite labeling criteria for a particular diagnosis, then it is assumed that this patient will respond to intervention in a manner resembling that previously demonstrated by other communities treated for this condition. Accurate diagnostic labeling gives the individual access to the total therapeutic knowledge base currently available within the health care system. The management options of the individual practitioner given a specific diagnostic label may be to treat, refer, or initiate team care for the patient. Regardless of the approach selected, it is important that the therapy be both effective and safe. Safety and efficacy are determined from controlled randomized clinical trials and longitudinal cohort studies. In the absence of such data, no definitive statement can be made on the safety or efficacy of a particular intervention. This statement holds true even if interventions appear credible on the grounds of being "natural" or having been persistently used by responsible professionals over many years.

The "natural is safe" dictum is flawed. Cortisol is a biological compound, yet therapy with pharmacological concentrations of steroids is associated with the development of numerous systemic complications including (25):

- Myopathy.
- Osteoporosis.
- Aseptic bone necrosis.
- Hypertension with sodium and water retention.
- Electrolyte distortion resulting in hypokalemic alkalosis.
- Mood changes ranging from euphoria to psychiatric disturbances.
- Pseudocerebral tumor.
- Peptic ulceration, sometimes leading to gastric hemorrhage and intestinal perforation.
- Pancreatitis.
- Impaired wound healing and atrophy of subcutaneous tissue, resulting from inhibition of fibroplasia.
- Secondary diabetes mellitus.
- Hyperlipidemia.
- Centripetal obesity.
- Glaucoma.
- Cataracts.
- Suppression of the immune system.
- Secondary amenorrhoea.
- Suppression of the hypothalamic–pituitary–adrenal axis. It is this complication that has necessitated development of an algorithm for corticosteroid withdrawal (26).

It may be argued that problems with corticosteroid usage relate to the dose rather than the substance used. Certainly, low-dose aspirin reduces the risk of side effects and appears to have a therapeutic platelet-inhibitory effect comparable to or greater than that of larger doses. Low-dose aspirin therapy is recommended for men over the age of 40 who are at significant risk for myocardial infarction and have no contraindications to this drug (27). Dose emerges as a pertinent variable in conducting safe clinical management. Animal experiments are used to determine the minimal lethal dose. A measure of drug safety can be ascertained by determining the difference between the dose required to achieve a therapeutic effect and that causing side effects. An optimal dose is one producing maximum benefit and minimum side effects. Ultimately, the optimal dose can only be accurately determined in clinical trials.

If natural composition is not a fail-safe guide to clinical intervention, then perhaps selection of intervention alternatives that have a long history of clinical use provides a safe approach? Aspirin, an over-the-counter drug with a prolonged history of clinical use, is associated with gastric erosions, occult and gross intestinal bleeding, tinnitus, and reversible deafness (28). Customary use is not synonymous with an inherently safe intervention or with the safe use of a particular intervention. Claims of safety based on traditional use and "naturalness" of an intervention are equally suspect, regardless of whether the proposed intervention targets biochemical or biomechanical body function.

Chiropractic clinical care has long been criticized for its lack of scientific validation. While manipulation is recognized as enjoying an extremely strong empirical track record (29), there is "little scientific data from which to evaluate this clinical approach to health and to the treatment of disease" (30). The importance of implementing clinical trials to establish a scientific basis to chiropractic clinical care is acknowledged (31). The consensus arising from a Delphi study cited the following as three of the five most important areas for future chiropractic research attention (32): controlled clinical trials for evaluating the use of spinal manipulative therapy; verification of the subluxation; and evaluation of the effectiveness of manipulation on physiological function. The conclusions of this study were derived from three mailings to 50 authoritative chiropractors, drawn from the ranks of American Chiropractic College presidents and ACA-approved councils and committees, and four nonchiropractic researchers with a track record of involvement with the chiropractic profession. The chiropractic profession recognizes that it is not sufficient to base the clinical practice of its members on the subjective improvement witnessed by thousands of chiropractors and millions of patients; it recognizes that chiropractic requires better substantiation of efficiency than clinical experience and anecdotal accounts (33). Despite lacking scientifically acceptable validation of its efficacy, chiropractic care is regarded as a fairly safe form of clinical intervention. This safety record may be partly attributable to the formulation of clear guidelines for contraindications to manipulations, following the recognition that manipulation is not necessarily a harmless procedure (34).

Contraindication to bony adjustment may be absolute when there are anatomical reasons that preclude safe manipulation, and relative in cases of functional patient disorders or individual practitioner incompetencies. Contraindications to manipulation include (35):

- Articular derangements, such as those occurring in the arthritides, in joint hypermobility, or following articular trauma such as whiplash.
- Bone weakening diseases; whether attributable to osteoporosis, osteomalacia, or malignancy, fragile bone constitutes an absolute contraindication to manipulation.
- Vascular accidents; aneurysms may be ruptured by spinal adjustments in an adjacent area.
- Neurological dysfunction, whether attributable to nerve root irritation in instances of disc prolapse or to functional intolerances such as vertigo on assuming various neck postures.
- Psychological and/or pain intolerance.
- Incomplete or inadequate diagnostic evaluation or work up.
- Poorly developed and/or inadequately maintained skill on the part of the practitioner.
- Poor technique implementation.
- An inadequate or outdated knowledge base.

A variety of accidents have been documented to occur with inappropriate cervical manipulation. These include vertebrobasilar insults, atlas compression of the internal carotid artery, hematomyelia due to cervical hyperextension, and dislocation of the atlas on the axis. Accidents involving manipulation of the cervical spine can be fatal. Warning signs of brainstem ischemia include (36):

- Vertigo and lightheadedness.
- Loss of consciousness.
- Visual disturbances.
- Nausea and vomiting.
- Incoordination of the extremities, difficulty walking, falling to one side.
- Numbness on one side of the body.
- Tinnitus.
- Speech difficulties.

Although nonfatal, manipulation accidents involving the lumbar spine can be most distressing. Attempted manipulation in instances of disc herniation and/or prolapse can lead to the development of the cauda equina syndrome, with pain and numbness over the buttocks and posterior thighs and extension of sensory changes down the backs of the legs to involve the feet. Muscular weakness is detected in the involved area. Loss of bladder and bowel control is an ominous sign. Cauda equina syndrome is a surgical emergency.

Safe therapy implies that the risk or cost of an intervention is low. Absolute safety is an important variable in selecting a particular intervention; it is, however, a less heavily weighted therapeutic consideration than effectiveness. The natural prognosis of the uninterrupted disease is another important consideration; when the outcome is early death or profound morbidity, then less safe but effective therapies are contemplated. Chloramphenicol carries the risk of bone marrow depression; nonetheless, it is the treatment of choice in the management of typhoid fever caused by ampicillin-resistant strains of *Salmonella typhi*. Complications occur in 30% of untreated cases and account for 75% of deaths in typhoid fever. Use of ampicillin in the management of ampicillin-resistant typhoid fever avoids any risk of the rare, potentially irreversible, and sometimes fatal agranulocytosis associated with chloramphenicol therapy; it also has no impact on the natural progression of typhoid fever. A safe therapy that fails to influence disease outcome or provide the patient with some tangible benefit is contraindicated. The cost–benefit ratio of an ineffective therapy approximates infinity. Some effective therapies may be contraindicated owing to their adverse side effects; all ineffective therapies are contraindicated despite being apparently "completely" safe. Some effective therapies may be indicated despite their side effects, given the prevalence of serious complications in untreated disease. Appropriate clinical management therefore implies careful consideration of both the indications and the contraindications to the use of any particular intervention strategy. Both indications and contraindications to therapy evolve from clinical practice experience. Validation of this experience is best achieved within the objective framework of randomized controlled clinical trials. It is only within such a framework that a clinical environment can be created in which three important confounding variables can be addressed. The placebo, Hawthorne, and Pygmalion effects all contribute to the perceived efficacy of an intervention provided by a confident, enthusiastic practitioner to the trusting, help-seeking patient (37).

"SAFE" KNOWLEDGE

One of the potential contraindications to manipulation is an ill-informed practitioner. A practitioner may become ill-informed as a result of an outdated knowledge base or a gullible and unquestioning acceptance of clinical information. The importance of continuing education is universally recognized amongst health professionals. There are two vital phases in any personal continuing education program. The first stage is to clearly identify what information is relevant to the individual's clinical practice; the second is to differentiate which knowledge and skills can justifiably be included in the practice situation.

The stages involved in determining personal professional learning needs include identifying: the optimal practice knowledge and skills that will meet the needs of those patients currently seeking assistance from the practitioner's clinic; areas of particular clinical interest to the practitioner; and trends that are likely to assume significance in chiropractic health care in the future (2). Once pertinent subject areas have been identified, the practitioner's task is to evaluate his or her current competence in these areas. The difference between the practitioner's self-perceived competence and the knowledge and skills required for optimal performance in the chosen area of clinical care may be defined as the practitioner's learning need. A learning need is the gap between the knowledge and skills required for optimal function and the current level of competence in an area. The perceived learning need is then converted into the learning objective.

Learning objectives are best stated in behavioral terms. A practitioner who decides to expand health promotion within the clinic may determine that patients would benefit from a smoking cessation program that goes beyond the current "stop smoking" directive. The practitioner's learning objective is to develop a stop smoking program that may be effectively used within a chiropractic clinic (38). The next stage is the identification of strategies whereby such an objective may be achieved. The immediate goal in this instance would be to perform a literature search to obtain knowledge about the current practice techniques used to encourage smoking cessation in clinical practice. In general, clarity, usefulness, comprehensiveness, reliability, and validity are the criteria used in knowledge selection.

A long-term aim in the achievement of this learning objective could be the implementation and evaluation of the selected smoking management program in the practitioner's own clinic. Monitoring for any changes in smoking habit provides a basis for evaluating the success of the learning objective in the practice situation. Success could be measured against the standard achieved by general medical practitioners. Primary practitioners have been found to favorably alter their patient's smoking habits (39); techniques are available for predicting which individuals are most likely to comply with such behavioral interventions (40).

In summary, continuing self-education passes cyclically through the phases of identifying a learning need, stating the learning need in terms of learning objectives, acquiring knowledge and skills to meet these learning objectives, and then evaluating learned information, against predetermined success criteria, in the performance of clinical tasks.

Basic to successful continuing self-education is the ability to gain access to current information. This may be achieved by regular attendance at professional meetings, workshops, and seminars or enrollment in formal continuing education courses; it may be addressed by regular reading of professional journals. Directed professional self-education frequently requires an in-depth literature review of areas designated as learning needs. With the communications revolution, access to a vast store of professional knowledge is possible (41). Of 15 computerized data bases of medical literature, Medline is the largest and most popular. Medline permits the reader to look up articles in a given year by author or medical subject heading terms; searches can be facilitated with the use of checklists and flowcharts (42). Paperchase is an alternative system that emphasizes bibliographic retrieval (43). The chiropractic literature is not yet comprehensively accessed on a computer network; Chiropractic Research Archives Collection (CRAC) provides a hard-copy listing of recent research work. The amount of health-relevant information is overwhelming. A strategy is required for selecting quality journal articles.

An algorithm has been proposed that takes advantage of the quality control already built into the normal publishing and writing habits of authors: the refereeing process of journal literature, the selection of experts to review articles, and the selection of refereed published documents by these experts in the preparation of their articles (44). The following three consecutive levels of quality filters are utilized:

- Identification of subject experts as evaluators. Reviews of particular areas are usually commissioned from recognized authorities in the area. It is helpful to peruse the publications of such authorities.
- Identification of the literature of a chosen subject. The bibliography of review articles provides a useful starting data base.
- Identification of key documents on the subject.

A number of potentially relevant journal articles can be selected using this strategy; further evaluation of each article is then required to determine whether it has relevance to the individual's particular learning need. This may be ascertained by, in the order listed (45):

- Examining the title to determine whether the topic is potentially relevant.
- Reviewing the list of authors referenced. This provides some substantiation to the basis for the article's conclusions.
- Reading the summary. This is to determine whether the results of the article, if correct, would be useful to the reader's clinical practice or to meeting a particular learning objective.
- Considering the circumstances in which the study was performed.

The relevance of published findings to a practitioner's clinical practice is a function of how readily the findings may be transferred and applied to the practitioner's patient population. When it appears that such information may be transferable because of the similarity of the study group and the practitioner's patient population, further critical analysis of the article is necessary. This includes evaluating several aspects of the publication.

1. The clinical data on which the article is based (46). Check for congruity between the various clinical findings reported, and note whether evidence and not merely inference has been documented.

2. The investigator's adherence to the basic rules underlying the chosen research methodology (47). Is sampling bias distorting findings? Were objective outcome criteria developed and specified? Was "blind" assessment performed? If new diagnostic tests are used, their relationship to the gold standard for diagnosing the condition should be reviewed. Intervention that succeeds in an incorrectly labeled disease is helpful only in treating the study patients, not the community with that disease label.

3. The relationship between the intervention and the outcome (48). A randomized trial is the best clinical research method currently available for determining a causal relationship. In evaluating an intervention, its relationship to any improvements and/or to new problems should be noted. Careful observation of clinical progress provides information about the effectiveness and safety of an intervention, given an appropriate research methodology. Before one can distinguish between interventions that may be useful, useless, or harmful in a particular practitioner's clinic, it is necessary to ascertain whether (22,49):

- The assignment of patients to groups was actually random. Is the comparison of clinical outcome affected by differences between the two groups and other prognostic variables?
- All clinically relevant outcomes are reported. Is the result influenced by errors in ascertaining the outcome?
- The study patients are recognizably similar to the practitioner's clinic patients. Are the results transferable?
- The clinical significance of the intervention was considered. Is this intervention preferable to that already available? What is the effect on the disease process; what are the side effects?
- The statistical significance of the intervention was considered. Are the results attributable to chance variation?
- The intervention is feasible in a chiropractic practice. Is this intervention compatible with the scope of chiropractic care and the individual practitioner's competence?
- All the patients who entered the study are accounted for at its conclusion. Has the outcome been distorted by unreported findings or by loss of patients from the trial?

Clinical trials do not pretend to offer irrefutable proof; they merely offer a standardized methodology whereby data may be collected and evaluated. Confounding variables that may distort results arising from clinical trials are corrected where possible (50). Investigator bias may be minimized by "blinding" investigators; patient bias may be addressed by providing a control group with sham therapy; the placebo effect may be controlled by experimental designs such as parallel groups or cross-over trials; bias in patient and control selection and in the allocation of intervention may be eliminated by randomization; and the probability that reported findings are due to chance can be ascertained by statistical analysis of the collected data.

All journal articles should be critically analyzed. Checklists to facilitate this process are available (51). Some provide a framework for assessing diagnostic tests; others provide an outline for evaluating the efficacy and effectiveness of therapeutic interventions (52,53). Checklists are even available for evaluating statistical information (54).

Continuing education is readily available to all. The value of this activity to clinical practice, however, is a function of the care with which the individual's learning need is identified and the critical analysis to which available "knowledge" is subjected. A commitment to continuing personal education and critical evaluation of the information encountered are essential prerequisites to safe contemporary practice.

Upon graduation, all chiropractors take an oath—the gist of which implies a commitment to prudent health care. Only diligent diagnostic work ups, fastidious management habits, and a healthy cynicism toward unproven intervention hypotheses and techniques can enable practitioners to meet the ethical obligations of their clinical practice.

References

1. Smith LS: Evaluation and management of muscle contraction headache. *Nurse Pract* 13:20–27, 1988.
2. Jamison JR: *Holistic Health Care.* Melbourne, PIT Press, 1984, pp 1–11, 79–112.
3. Edmeads J: The worse headache ever: ominous causes. *Postgrad Med* 86:93–104, 1989.
4. Edmeads J: Headache in cerebrovascular disease. *Postgrad Med* 81:191–197, 1987.
5. Becker L, Iverson DC, Reed FM, Calogne N, Miller RS, Freeman WL: Patients with new headache in primary care: a report from ASPN. *J Fam Pract* 27:41–47, 1988.
6. Anthony M: Cluster headache: understanding the problem. *Patient Manage* Nov:51–59, 1984.
7. Selby G: Migraine, everybody's headache I: clinical features and pathogenesis. *Patient Manage* Nov:29–41, 1984.
8. Diamond S, Prager J, Freitag FG: Diet and headache. *Postgrad Med* 79:279–285, 1986.
9. Perkin JE, Hartje J: Diet and migraine: a review of the literature. *J Am Diet Assoc* 83:459–463, 1983.
10. Lance JW: The pharmacology of migraine. *Med J Aust* 144:85–88, 1986.
11. Arthur GP: Migraine, everybody's headache II: treatment. *Patient Manage* Nov:43–49, 1984.
12. Elkind AH: Muscle contraction headache. *Postgrad Med* 81:203–217, 1987.
13. McKenna JP: Cluster headache. *Am Fam Physician* 37:173–178, 1988.
14. Sutherland JE: Headaches. *Am Fam Physician* 27:137–142, 1983.
15. Spira PJ: Migraine vs tension headache. *Patient Manage* May:115, 1987.
16. Vernon H: Chiropractic manipulative therapy in the treatment of headaches: a retrospective and prospective study. *J Manipulative Physiol Ther* 5:109–112, 1982.
17. Brooks PM: Musculoskeletal causes of chest pain. *Patient Manage* June:51–59, 1986.
18. Bennet RM, Smythe HA, Wolfe F: Recognizing fibromyalgia. *Patient Care* July:60–83, 1989.
19. Seymour J: Ischaemic cardiac pain. *Aust Fam Physician* 13:327–329, 1984.

20. Diamond GA, Forrester JS: Analysis of probability as an aid in the clinical diagnosis of coronary artery disease. *N Engl J Med* 300:1350, 1979.
21. Burstow D, Murphy A: The recognition and early management of acute myocardial infarction. *Aust Fam Physician* 13:319-324, 1984.
22. Sackett DL, Haynes RB, Tugwell P: *Clinical Epidemiology: A Basic Science for Clinical Medicine.* Boston, Little, Brown & Co, 1985, pp 59-138, 285-352.
23. Bett JHN, Davies P: Non ischaemic chest pain. *Aust Fam Physician* 13:336-340, 1984.
24. Garrett J: When the lung causes chest pain. *Patient Manage* June:60-75, 1986.
25. Melby JC: Drug spotlight program: systemic corticosteroid therapy: pharmacological and endocrinological consideration. *Ann Intern Med* 81:505-512, 1974.
26. Kountz DS: An algorithm for corticosteroid withdrawal. *Am Fam Physician* 39:250-258, 1989.
27. US Preventive Services Task Force. Aspirin prophylaxis for cardiovascular disease. *Am Fam Physician* 40:117-120, 1989.
28. Pinals RS: Rheumatoid arthritis: a pharmacologic overview. *Am Fam Physician* 37:145-152, 1988.
29. Hildebrandt RW: The research status of spinal manipulative therapeutics. *J Manipulative Physiol Ther* 1:221-222, 1978.
30. Hildebrandt RW: Research in spinal manipulation—reflections on processes, priorities and responsibilities. *J Manipulative Physiol Ther* 2:138, 1979.
31. Haldeman S: Basic principles for establishing a chiropractic clinical trial. *J Can Chiropractic Assoc* 12:S33-S37, 1978.
32. Phillips RB: Where do we go from here? The future of chiropractic research. *ACA Chiropractic* 20:26-28, 1983.
33. De Boer K, Waagen G: The future role of the chiropractor in the health care system. *J Manipulative Physiol Ther* 9:225-228, 1986.
34. Kleynhans AM, Terrett AGJ: The prevention of complications from spinal manipulative therapy. In Glasgow EF, et al (eds): *Aspects of Manipulative Therapy.* London, Churchill Livingstone, 1985, pp 161-175.
35. Kleynhans AM: Complications of and contraindications to spinal manipulative therapy. In Haldeman S (ed): *Modern Developments in the Principles and Practice of Chiropractic.* New York, Appleton-Century-Crofts, 1980, pp 359-384.
36. Terrett AJ: Vascular accidents from cervical spine manipulation: the mechanisms. *J Aust Chiropractors Assoc* 17:131-144, 1987.
37. Phillips RB: The challenge of proving the efficacy of chiropractic, placebo, Hawthorne and Pygmalion effects in research. *ACA Chiropractic* 20:30-40, 1983.
38. Jamison JR: *The Practitioner's Handbook of Disease Intervention: Part III: The Respiratory System.* Melbourne, Bookpit, 1984, pp 10-17.
39. Russel M, et al: Effect of general practitioner advice against smoking. *Br Med J* 2:231-235, 1979.
40. Jamison JR: Locus of control—an aid to clinical care. *J Aust Chiropractors Assoc* 17:25-27, 1987.
41. Brown JH: The communications revolution. *Hosp Forum* 22:21-24, 53-54, 1979.
42. Kolner SJ: Improving the MEDLARS search interview: a checklist approach. *Bull Med Libr Assoc* 69:26-33, 1981.
43. Horowitz GL, Bleich HL: Paperchase: a computer program to search the medical literature. *N Engl J Med* 305:924-930, 1981.
44. Pao ML: A quality filtering system for medical literature. *J Med Educ* 50:353-359, 1975.
45. Department of Clinical Epidemiology and Biostatistics, McMaster University Health Sciences Centre: How to read clinical journals: I. Why to read them and how to start reading them critically. *Can Med Assoc J* 124:555-590, 1981.
46. Department of Clinical Epidemiology and Biostatistics, McMaster University Health Sciences Centre: Clinical disagreement: II. How to avoid it and how to learn from one's mistakes. *Can Med Assoc J* 123:613-617, 1980.
47. Department of Clinical Epidemiology and Biostatistics, McMaster University Health Sciences Centre: How to read clinical journals: III. To learn the clinical course and prognosis of disease. *Can Med Assoc J* 124:869-872, 1981.

48. Department of Clinical Epidemiology and Biostatistics, McMaster University Health Sciences Centre: How to read clinical journals: IV. To determine etiology and causation. *Can Med Assoc J* 124:985–1031, 1981.
49. Elwood JM: Interpreting clinical trial results: seven steps to understanding. *Can Med Assoc J* 123:343–345, 1980.
50. Ramsay LE: Value and limitations of controlled clinical trials. *Med Int* 7:286–289, 1984.
51. Krogh CL: A checklist system for critical review of medical literature. *Med Educ* 19:392–395, 1985.
52. Nachemson AL, Larocca H: Editorial. Spine 1987. *Spine* 12:427–429, 1987.
53. Lionel NDW, Herxheimer A: Assessing reports of therapeutic trials. *Br Med J* 3:637–640, 1979.
54. Berry G: Statistical guide-lines and statistical guidance. *Med J Aust* 146:408–409, 1987.

Index

Biomedical Model, 4
Bone Metabolism, 58–62

Calcitonin, 59–60
Calcium Absorption, 65
Calcium Deficiency Hypothesis, 62–69
Calcium Excretion, 65–67
Calcium Homeostasis, 63
Causal Diagnosis, 55–87
Clinical Decision Making, 1–15
Cluster Headache, 114
Collagen, 59
Consensus Conferences, 2

Diagnostic Labeling, 89–109
Dietary Calcium Adequacy, 64–65
Disease Classifications, 6–7
Dualism, 3

Epidemiology, 11
Estrogen Deprivation, 69–70

Fibromyalgia, 117–118
Fluoride Therapy, 81–82

Gouty Arthritis Signs, 41

Headache, 112–119
Health Examinations, 21–24
Holistic Model, 5
Hormone Replacement Therapy, 80–81
Hypertension, Diagnostic Decisions, 90–94
Hypertension, Management Decision Criteria, 94–105

Joint Pain, 37–42

Likelihood Ratio, 27–31

Meningitis, 113
Migraine Headache, 114

Osteitis Fibrosa Cystica, 56
Osteomalacia, 56
Osteoporosis, 56–58
Osteoporosis, Assessment, 71–74
Osteoporosis, Calcium Therapy, 77–79
Osteoporosis, Etiology, 62–70
Osteoporosis, Exercise Therapy, 75–77

Pattern Recognition, 37–54
Phosphorous and Magnesium Balance, 68
Popper, Karl, 3
Probability Reasoning, 17–36

Randomized Clinical Trial, 9–10
Reductionism, 3
Rheumatoid Arthritis, 45–47
Rheumatoid Arthritis Signs, 39–40
Rheumatoid Arthritis, Dietary Management, 47–50
Risk Factors, 18–19

Safe Diagnosis, 111–119
Safe Health Care, 111–128
Safe Knowledge, 123–126
Safe Management, 119–123
Subarachnoid Hemmorhage, 112–113

Tension Headache, 115–116

Vitamin D, 60–62